PE

civil engineering

construction

practice exam

NCEES
advancing licensure for
engineers and surveyors

978-1-932613-70-4

ISBN 978-1-932613-70-4

Printed in the United States of America
October 2014 First Printing

CONTENTS

About NCEES

The National Council of Examiners for Engineering and Surveying (NCEES) is a nonprofit organization made up of engineering and surveying licensing boards from all U.S. states and territories. Since its founding in 1920, NCEES has been committed to advancing licensure for engineers and surveyors in order to protect the health, safety, and welfare of the American public.

NCEES helps its member licensing boards carry out their duties to regulate the professions of engineering and surveying. It develops best-practice models for state licensure laws and regulations and promotes uniformity among the states. It develops and administers the exams used for engineering and surveying licensure throughout the country. It also provides services to help licensed engineers and surveyors practice their professions in other U.S. states and territories.

Updates on exam content and procedures

Visit us at **ncees.org/exams** for updates on everything exam-related, including specifications, exam-day policies, scoring, and corrections to published exam preparation materials. This is also where you will register for the exam and find additional steps you should follow in your state to be approved for the exam.

Exam-day schedule

Be sure to arrive at the exam site on time. Late-arriving examinees will not be allowed into the exam room once the proctor has begun to read the exam script. The report time for the exam will be printed on your Exam Authorization. Normally, you will be given 1 hour between morning and afternoon sessions.

Admission to the exam site

To be admitted to the exam, you must bring two items: (1) your Exam Authorization and (2) a current, signed, government-issued identification.

Examinee Guide

The *NCEES Examinee Guide* is the official guide to policies and procedures for all NCEES exams. All examinees are required to read this document before starting the exam registration process. You can download it at ncees.org/exams. It is your responsibility to make sure that you have the current version.

NCEES exams are administered in either a computer-based format or a pencil-and-paper format. Each method of administration has specific rules. This guide describes the rules for each exam format. Refer to the appropriate section for your exam.

Scoring and reporting

NCEES typically releases exam results to its member licensing boards 8–10 weeks after the exam. Depending on your state, you will be notified of your exam result online through your MyNCEES account or via postal mail from your state licensing board. Detailed information on the scoring process can be found at ncees.org/exams.

Staying connected

To keep up to date with NCEES announcements, events, and activities, connect with us on your preferred social media network.

EXAM SPECIFICATIONS

NCEES Principles and Practice of Engineering
CIVIL BREADTH and CONSTRUCTION DEPTH
Exam Specifications
Effective Beginning with the April 2015 Examinations

- The civil exam is a breadth and depth examination. This means that examinees work the breadth (AM) exam and one of the five depth (PM) exams.

- The five areas covered in the civil exam are construction, geotechnical, structural, transportation, and water resources and environmental. The breadth exam contains questions from all five areas of civil engineering. The depth exams focus more closely on a single area of practice in civil engineering.

- Examinees work all questions in the morning session and all questions in the afternoon module they have chosen. Depth results are combined with breadth results for final score.

- The exam is an 8-hour open-book exam. It contains 40 multiple-choice questions in the 4-hour AM session, and 40 multiple-choice questions in the 4-hour PM session.

- The exam uses both the International System of Units (SI) and the US Customary System (USCS).

- The exam is developed with questions that will require a variety of approaches and methodologies, including design, analysis, and application.

- The knowledge areas specified as examples of kinds of knowledge are not exclusive or exhaustive categories.

- The specifications for the **AM exam** and the **Construction PM exam** are included here. The **design standards** applicable to the Construction PM exam are shown on **ncees.org**.

CIVIL BREADTH Exam Specifications

	Approximate Number of Questions
I. Project Planning	**4**
A. Quantity take-off methods	
B. Cost estimating	
C. Project schedules	
D. Activity identification and sequencing	
II. Means and Methods	**3**
A. Construction loads	
B. Construction methods	
C. Temporary structures and facilities	
III. Soil Mechanics	**6**
A. Lateral earth pressure	
B. Soil consolidation	
C. Effective and total stresses	
D. Bearing capacity	
E. Foundation settlement	
F. Slope stability	

IV. Structural Mechanics   **6**
- A. Dead and live loads
- B. Trusses
- C. Bending (e.g., moments and stresses)
- D. Shear (e.g., forces and stresses)
- E. Axial (e.g., forces and stresses)
- F. Combined stresses
- G. Deflection
- H. Beams
- I. Columns
- J. Slabs
- K. Footings
- L. Retaining walls

V. Hydraulics and Hydrology   **7**
- A. Open-channel flow
- B. Stormwater collection and drainage (e.g., culvert, stormwater inlets, gutter flow, street flow, storm sewer pipes)
- C. Storm characteristics (e.g., storm frequency, rainfall measurement and distribution)
- D. Runoff analysis (e.g., Rational and SCS/NRCS methods, hydrographic application, runoff time of concentration)
- E. Detention/retention ponds
- F. Pressure conduit (e.g., single pipe, force mains, Hazen-Williams, Darcy-Weisbach, major and minor losses)
- G. Energy and/or continuity equation (e.g., Bernoulli)

VI. Geometrics   **3**
- A. Basic circular curve elements (e.g., middle ordinate, length, chord, radius)
- B. Basic vertical curve elements
- C. Traffic volume (e.g., vehicle mix, flow, and speed)

VII. Materials   **6**
- A. Soil classification and boring log interpretation
- B. Soil properties (e.g., strength, permeability, compressibility, phase relationships)
- C. Concrete (e.g., nonreinforced, reinforced)
- D. Structural steel
- E. Material test methods and specification conformance
- F. Compaction

VIII. **Site Development**   **5**
    A. Excavation and embankment (e.g., cut and fill)
    B. Construction site layout and control
    C. Temporary and permanent soil erosion and sediment control (e.g., construction erosion control and permits, sediment transport, channel/outlet protection)
    D. Impact of construction on adjacent facilities
    E. Safety (e.g., construction, roadside, work zone)

CIVIL–CONSTRUCTION DEPTH Exam Specifications

**Approximate Number
of Questions**

I. Earthwork Construction and Layout   6
   A. Excavation and embankment (e.g., cut and fill)
   B. Borrow pit volumes
   C. Site layout and control
   D. Earthwork mass diagrams and haul distance
   E. Site and subsurface investigations

II. Estimating Quantities and Costs   6
   A. Quantity take-off methods
   B. Cost estimating
   C. Cost analysis for resource selection
   D. Work measurement and productivity

III. Construction Operations and Methods   7
   A. Lifting and rigging
   B. Crane stability
   C. Dewatering and pumping
   D. Equipment operations (e.g., selection, production, economics)
   E. Deep foundation installation

IV. Scheduling   5
   A. Construction sequencing
   B. Activity time analysis
   C. Critical path method (CPM) network analysis
   D. Resource scheduling and leveling
   E. Time-cost trade-off

V. Material Quality Control and Production   6
   A. Material properties and testing (e.g., soils, concrete, asphalt)
   B. Weld and bolt installation
   C. Quality control process (QA/QC)
   D. Concrete proportioning and placement
   E. Concrete maturity and early strength evaluation

VI. Temporary Structures   **7**
    A. Construction loads, codes, and standards
    B. Formwork
    C. Falsework and scaffolding
    D. Shoring and reshoring
    E. Bracing and anchorage for stability
    F. Temporary support of excavation

VII. Health and Safety   **3**
    A. OSHA regulations and hazard identification/abatement
    B. Safety management and statistics
    C. Work zone and public safety

101. A 227-ft length of canal is to be lined with concrete for erosion control. With 12% allowance for waste and overexcavation, the volume (yd^3) of concrete that must be delivered is most nearly:

(A)   234
(B)   280
(C)   292
(D)   327

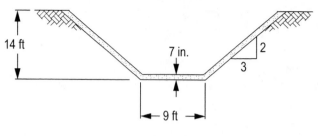

LINED LENGTH = 227 ft

102. Based on the straight-line method of depreciation, the book value at the end of the 8th year for a track loader having an initial cost of $75,000, and a salvage value of $10,000 at the end of its expected life of 10 years is most nearly:

(A) $10,000
(B) $15,000
(C) $23,000
(D) $48,750

103. The budgeted labor amount for an excavation task is $4,000. The hourly labor cost is $50 per worker, and the workday is 8 hours. Two workers are assigned to excavate the material. The time (days) available for the workers to complete this task is most nearly:

(A)   3
(B)   4
(C)   5
(D)   12.5

104. A CPM arrow diagram is shown below. Nine activities have been estimated with durations ranging from 5 to 35 days. The minimum time (days) required to finish the project is most nearly:

(A)   40
(B)   42
(C)   45
(D)   50

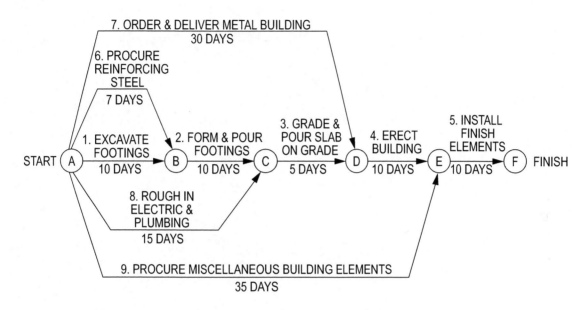

105. A bridge is to be jacked up to replace its bearings. The design requires a hydraulic ram with a minimum capacity of 1,000 kN (kilonewtons). The hydraulic rams that are available are rated in tons (2,000 lb/ton). The **minimum** size (tons) ram to use is most nearly:

(A)   1,110
(B)   250
(C)   150
(D)   100

106. A crane with a 100-ft boom is being used to set a small load on the roof of the building shown. The minimum standoff (Point A) from the corner of the building to the centerline of the boom is indicated. What is the maximum distance (ft) from the edge of the building that the load can be placed on the roof?

(A) 16
(B) 25
(C) 30
(D) 36

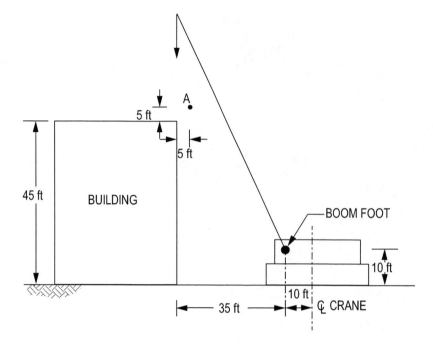

107. A wall form subjected to a wind load of 20 psf is prevented from overturning by diagonal braces spaced at 8 ft on center along the length of the wall form as shown in the figure. The connection at the base of the form at Point A is equivalent to a hinge. Ignore the weight of the form. The axial force (lb) resisted by the brace is most nearly:

(A) 2,050
(B) 2,560
(C) 2,900
(D) 4,525

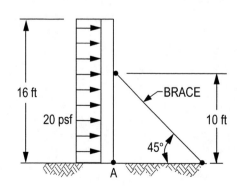

108. Which one of the following statements regarding lateral earth pressures is correct?

(A) The lateral strain required to fully mobilize the soil passive pressure is considerably smaller than the lateral strain required to fully mobilize the soil active pressure.

(B) The lateral strain required to fully mobilize the soil passive pressure is slightly smaller than the lateral strain required to fully mobilize the soil active pressure.

(C) The lateral strain required to fully mobilize the soil passive pressure is slightly greater than the lateral strain required to fully mobilize the soil active pressure.

(D) The lateral strain required to fully mobilize the soil passive pressure is considerably greater than the lateral strain required to fully mobilize the soil active pressure.

109. Site preparation and grading require the placement of 20 ft of new fill. An analysis of the resulting consolidation of the underlying soft, saturated, compressible deposits reveals a mean consolidation settlement of 22 in. affecting a 21.5-acre area. Prefabricated wick drains will be used to accelerate the settlement to meet the project schedule. Because of contamination from the former site use, the effluent from the wick drains will need to be collected and treated prior to disposal at an estimated cost of $0.25 per gallon. Assuming no loss of effluent during collection, the estimated treatment and disposal cost for the wick drain effluent at this site is most nearly:

(A) $430,000
(B) $3,200,000
(C) $5,200,000
(D) $35,000,000

110. A soil profile is shown in the figure. The effective vertical stress (psf) at Point A is most nearly:

(A)   1,270
(B)   1,820
(C)   2,140
(D)   2,570

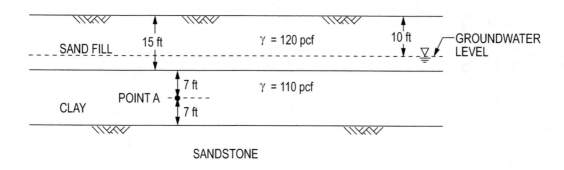

111. A bridge footing is to be constructed in sand. The groundwater level is at the ground surface. The ultimate bearing capacity would be based on what type of soil unit weight?

(A)   Buoyant unit weight

(B)   Saturated unit weight

(C)   Dry unit weight

(D)   Total unit weight

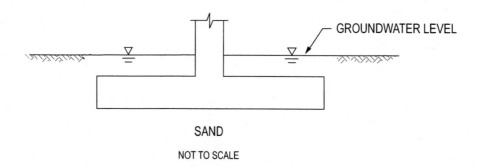

112. The figure shows two identical building footings with the same load but constructed in two different soil types. Which of the following statements is most correct?

(A)   The long-term settlement for Case I is less than Case II.

(B)   The long-term settlement for Case II is less than Case I.

(C)   The long-term settlements are the same for both cases.

(D)   Settlement is not a concern for either case.

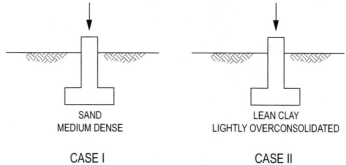

113. The minimum factor of safety against rotational failure for permanent slopes under long-term, non-seismic conditions influencing occupied structures is closest to:

(A)   1.0
(B)   1.1
(C)   1.5
(D)   3.0

114. Referring to the figure, what load combination produces the maximum uplift on Footing A?

(A)   Dead + live

(B)   Dead + wind

(C)   Dead

(D)   Dead + live + wind

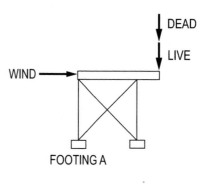

115. A simply supported truss is loaded as shown in the figure. The loads (kips) for Members b and c are most nearly:

	Member b	**Member c**
(A)	0	0
(B)	0	100
(C)	100	0
(D)	100	100

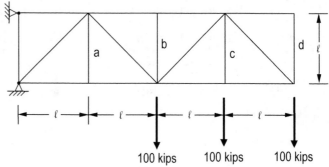

116. Consider two beams with equal cross-sections, made of the same material, having the same support conditions, and each loaded with equal uniform load per length. One beam is twice as long as the other. The maximum bending stress in the longer beam is larger by a factor of:

(A)   1.25
(B)   2
(C)   3
(D)   4

117. The point load (kips) placed at the centerline of a 30-ft beam that produces the same maximum shear in the beam as a uniform load of 1 kip/ft is most nearly:

(A)   7.5
(B)   15
(C)   30
(D)   60

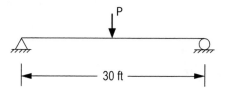

118. The beam sections shown are fabricated from 1/2-in. × 6-in. steel plates. Which of the following cross sections will provide the greatest flexural rigidity about the x-axis?

(A)

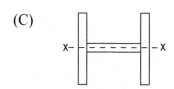

(B)

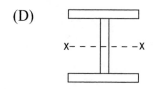

(C)

(D)

119. A concrete gravity retaining wall having a unit weight of 150 pcf is shown in the figure. Use the Rankine active earth pressure theory and neglect wall friction. The factor of safety against overturning about the toe at Point O is most nearly:

(A) 3.1
(B) 2.5
(C) 2.2
(D) 0.3

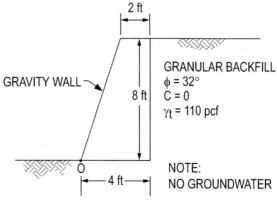

120. A drainage basin produces a stormwater runoff volume of 25.0 acre-ft, which must be drained through a rectangular channel that is 4 ft wide and 2 ft deep and has a uniform slope of 0.2%. Assume a Manning roughness coefficient of 0.022 and a constant depth of flow of 1.5 ft. The time (hours) it will take to discharge the runoff is most nearly:

(A) 12.5
(B) 16.4
(C) 18.5
(D) 25.0

121. Two identical 12-in. storm sewers flow full at a 2% slope into a junction box. A single larger pipe of the same material and slope flows out of the box. Assuming the following pipe sizes are commercially available, the minimum size of this downstream pipe (in.) designed to flow full is most nearly:

(A)   16
(B)   18
(C)   20
(D)   24

122. The following table represents the rainfall recorded from all rain gages located in and around a drainage area.

Gage	A	B	C	D	E	F	G	H	I	J	K
Rainfall (in.)	2.1	3.6	1.3	1.5	2.6	6.1	5.1	4.8	4.1	2.8	3.0

Using the arithmetic mean method, the average precipitation (in.) for the drainage area is most nearly:

(A)   3.4
(B)   3.7
(C)   4.1
(D)   37.0

123. The rational method must be used to determine the maximum runoff rate for a 90-acre downtown area. The time of concentration for the 50-year frequency storm is 1 hour. Intensity-duration-frequency curves and a table of runoff coefficients are provided. The maximum runoff rate (cfs), based on the maximum runoff coefficient for a 50-year storm, is most nearly:

(A) 160
(B) 220
(C) 300
(D) 340

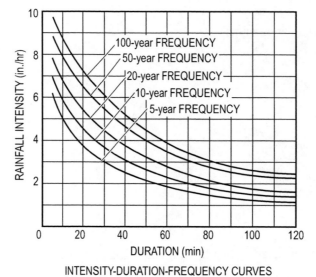

INTENSITY-DURATION-FREQUENCY CURVES

123. (Continued)

Description of Area	Runoff Coefficients
Business	
Downtown areas	0.70–0.95
Neighborhood areas	0.50–0.70
Residential	
Single-family areas	0.30–0.50
Multiunits, detached	0.40–0.60
Multiunits, attached	0.60–0.75
Residential (suburban)	0.25–0.40
Apartment dwelling areas	0.50–0.70
Industrial	
Light areas	0.50–0.80
Heavy areas	0.60–0.90
Parks, cemeteries	0.10–0.25
Playgrounds	0.20–0.35
Railroad yard areas	0.20–0.40
Unimproved areas	0.10–0.30
Streets	
Asphalt	0.70–0.95
Concrete	0.80–0.95
Brick	0.70–0.85
Drives and walks	0.75–0.85

124. A stormwater drainage ditch with a maximum capacity of 10 cfs discharges into a detention basin. The detention basin volume is 400,000 gal. During a storm event the average discharge into the detention basin was 1.5 cfs. The time (hours) to fill the empty basin would be most nearly:

(A) 1.5
(B) 9.9
(C) 11.1
(D) 74.1

125. Assume fully turbulent flow in a 1,650-ft section of 3-ft-diameter pipe. The Darcy-Weisbach friction factor f is 0.0115. There is a 5-ft drop in the energy grade line over the section. The flow rate (cfs) is most nearly:

(A) 16
(B) 29
(C) 50
(D) 810

126. Assuming that Bernoulli's equation applies (ignore head losses) to the pipe flow shown in the figure, which of the following statements is most correct?

(A) Pressure head increases from 1 to 2.

(B) Pressure head decreases from 1 to 2.

(C) Pressure head remains unchanged from 1 to 2.

(D) Bernoulli's equation does not include pressure head.

DIRECTION OF FLOW

127. The following information is for a proposed horizontal curve in a new subdivision:

PI station	12+40.00
Degree of curve	10°
Deflection angle	12°30′

The station of the PT is most nearly:

(A) 12+79.80
(B) 12+80.10
(C) 13+02.00
(D) 13+64.75

128. For the sag vertical curve shown, the tangent slope at Station 14+00 is most nearly:

(A) +0.53%
(B) +1.23%
(C) +2.12%
(D) +2.77%

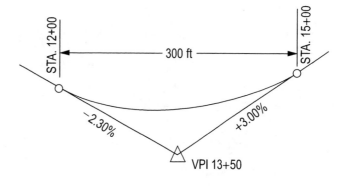

NOT TO SCALE

129. An interstate highway has the following traffic count data for a day in each month as shown below:

Jan.	63,500
Feb.	62,100
Mar.	64,400
Apr.	64,900
May	75,800
June	77,300
July	78,950
Aug.	77,200
Sept.	70,050
Oct.	69,000
Nov.	66,000
Dec.	64,000
Annual Total	833,200

The seasonal factor for the summer months of June through August is most nearly:

(A) 0.28
(B) 0.89
(C) 1.02
(D) 1.12

130. The most essential criteria for proper soil classification using the Unified Soil Classification or the AASHTO Soil Classification system are:

(A) water content and soil density

(B) Atterberg limits and specific gravity

(C) grain-size distribution and water content

(D) grain-size distribution and Atterberg limits

131. The Standard Penetration Test (SPT) is widely used as a simple and economic means of obtaining which of the following?

    (A)   A measurement of soil compressibility expressed in terms of a compression index

    (B)   A direct measurement of the undrained shear strength

    (C)   An indirect indication of the relative density of cohesionless soils

    (D)   A direct measurement of the angle of internal friction

132. A department of transportation must remove and replace a 12-ft × 20-ft concrete slab on an interstate facility. To minimize disruption to traffic, the work must be completed during an 8-hour nighttime work shift. Nighttime temperatures average 50°F. If the minimum required compressive strength is 3,500 psi, the concrete mix would most likely consist of:

    (A)   coarse aggregate, sand, Type II cement, chemical accelerator

    (B)   sand, Type III cement, water, chemical accelerator

    (C)   coarse aggregate, sand, Type V cement, water, chemical accelerator

    (D)   coarse aggregate, sand, Type III cement, water, chemical accelerator

133. Fatigue in steel can be the result of:

    (A)   a reduction in strength due to cyclical loads

    (B)   deformation under impact loads

    (C)   deflection due to overload

    (D)   expansion due to corrosion

134. Sample concrete cylinders that are 6 inches in diameter and 12 inches high are tested to determine the compressive strength of the concrete f_c'. The test results are as follows:

Sample	Axial Compressive Failure Load (lb)
1	65,447
2	63,617
3	79,168

Based on the above results, the average 28-day compressive strength (psi) is most nearly:

(A)   615
(B) 2,250
(C) 2,450
(D) 2,800

135. During testing of a sample in the laboratory, the following soil data were collected:

Combined weight of compacted soil sample and the mold is 9.11 lb.

Water content of soil sample is 11.5%.

The weight and volume of mold are 4.41 lb and 0.03 ft³, respectively.

The dry unit weight of the soil sample (pcf) is most nearly:

(A)   160
(B)   140
(C)   127
(D)   125

28

136. Refer to the figure. The net excess excavated material (yd^3) from Station 1+00 to Station 3+00 is most nearly:

(A)   160
(B)   262
(C)   390
(D)   463

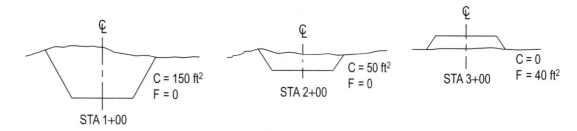

137. An existing pipe connects two maintenance holes (MH). A third MH is planned between the two. At the new MH, the elevation (ft) of the top of the pipe is most nearly:

(A)   623.06
(B)   627.56
(C)   628.06
(D)   628.56

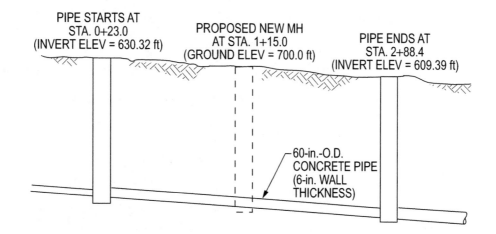

138. Which of the following is **not** a stormwater erosion classification?

    (A)   Sheet erosion

    (B)   Rill erosion

    (C)   Gully erosion

    (D)   Rushing erosion

139. Based on the soil classification system found in the federal OSHA regulation Subpart P, Excavations, the soil adjacent to an existing building has been classified as Type B. An undisturbed perimeter strip that is 5 ft wide is to be maintained along the face of the building. The excavation is to be 12 ft deep. To meet OSHA excavation requirements, the minimum horizontal distance X (ft) from the toe of the slope to the face of the structure is most nearly:

    (A)   11
    (B)   14
    (C)   17
    (D)   23

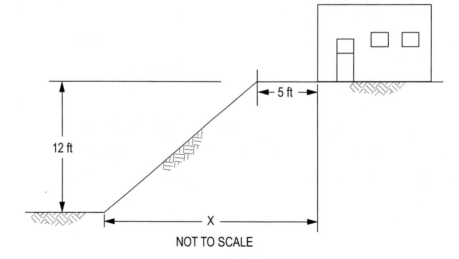

NOT TO SCALE

140. Based on the criteria provided, the steepest backslope (H:V) preferred in the ditch shown is most nearly:

(A)   2:1

(B)   3:1

(C)   5:1

(D)   6:1

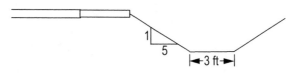

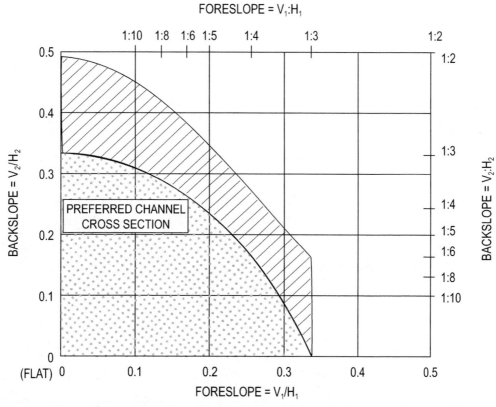

This area is applicable to all Vee ditches, rounded channels with a bottom width less than 2.4 m [8 ft], and trapezoidal channels with bottom widths less than 1.2 m [4 ft].

This area is applicable to rounded channels with bottom width of 2.4 m [8 ft] or more and to trapezoidal channels with bottom widths equal to or greater than 1.2 m [4 ft].

Adapted from AASHTO *Roadside Design Guide,* 4th edition, 2011.

This completes the morning session. Solutions begin on page 63.

501. A segment of interstate highway requires the construction of an embankment of 500,000 yd³. The embankment fill is to be compacted to a minimum of 90% of Modified Proctor maximum dry density.

A source of suitable borrow has been located for construction of the embankment. Assume that there is no soil loss in transporting the soil from the borrow pit to the embankment.

The following data apply:

Dry unit weight of soil in borrow pit	113.0 pcf
Moisture content in borrow pit	16.0%
Specific gravity of the soil particles	2.65
Modified Proctor optimum moisture content	13.0%
Modified Proctor maximum dry density	120.0 pcf

Assuming each truck holds 5.0 yd³ and the void ratio of the soil is 1.30 during transport, the **minimum** number of truckloads of soil from the borrow pit that are required to construct the embankment is most nearly:

(A) 100,000
(B) 150,000
(C) 200,000
(D) 250,000

502. The figure shows a survey grid for a borrow pit excavation. The number at each grid intersection is the depth of cut at that location. The total excavation volume (yd³) is most nearly:

(A) 41
(B) 61
(C) 184
(D) 246

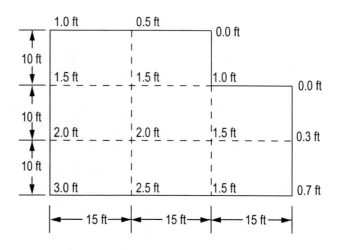

503. The curve shown in the figure has a radius R = 200 ft, and the mid-ordinate M = 12.8 ft. The length (ft) of the curve is most nearly:

(A) 143.9
(B) 140.8
(C) 75.2
(D) 71.9

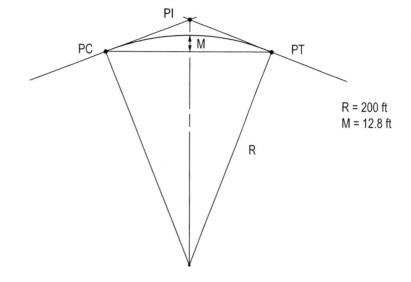

504. Referring to the grade profile and mass diagram for a roadway construction project, which of the following is/are true?

    I. The job is balanced (i.e., equal cut and fill).

    II. Section B–D represents a fill operation.

    III. Station D represents a transition point between cut and fill.

(A)   I only

(B)   II only

(C)   III only

(D)   II and III only

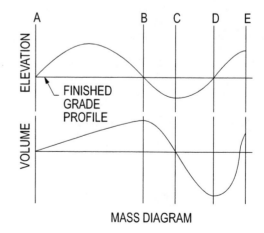

505. The geotechnical report for a project includes test borings and laboratory testing for the project area and potential borrow locations. The test borings include Standard Penetration Test (SPT) blowcounts and driven ring samples (with densities and moisture contents). Laboratory test results include gradations, standard Proctors (ASTM D698), sulfates, and shear strengths. What can be estimated from these results?

(A) Neither shrink nor swell

(B) Both shrink and swell

(C) Shrink can be estimated, but not swell

(D) Swell can be estimated, but not shrink

506. For the mass diagram shown, the total material (bcy) that must be moved along the project is most nearly:

(A) 6,000
(B) 5,000
(C) 3,000
(D) 2,000

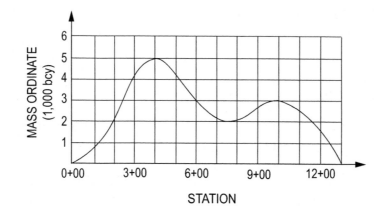

507. A reinforcing-steel quantity take-off has been completed for a 30-ft-long wall footing with the cross section shown. You have been asked to check the work. The total weight of reinforcing steel (lb) required for the footing is most nearly:

(A)   2,368
(B)   2,438
(C)   2,634
(D)   2,864

Quantity Take-off to Check				
Item	Number	Length (ft)	Unit Weight (lb/ft)	Weight (lb)
Transverse #6	60	5.5	1.502	495.66
Longit #4	14	29.5	0.668	275.88
BF Wall #7	119	4.5	2.044	1,094.56
FF Wall #4	30	3.83	0.668	767.53
Total weight				2,633.60

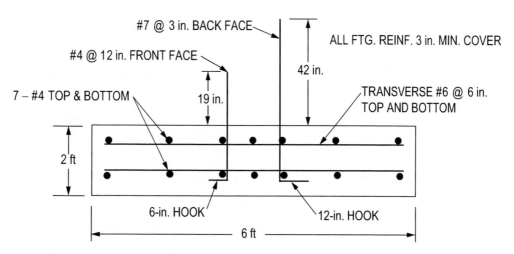

Bar Size	Weight (lb/ft)	Area (in²)
#3	0.376	0.11
#4	0.668	0.20
#5	1.043	0.31
#6	1.502	0.44
#7	2.044	0.60
#8	2.670	0.79
#9	3.400	1.00

508. A construction crew of six can form 48 ft^2 of contact area per crew-hour when working at less than 8 ft above grade. At more than 8 ft above grade, the production rate drops 40%. Assuming a crew cost of $372/hour, the labor cost to form walls for a 40-ft-long × 90-ft-wide × 14-ft-tall freestanding tank is most nearly:

    (A) $36,270
    (B) $56,420
    (C) $61,256
    (D) $72,540

509. A building requires 48 concrete beams that are 12 in. wide × 24 in. exposed depth × 30 ft long. The table indicates the amount of material required per 100 SFCA of concrete beams.

Number of Uses	Plywood (sf)	Lumber (bf)	Form Ties (ea)	Nails (lb)
1	105	275	12	3
2	50	140	12	2
3	40	110	12	1.5

A supplier has quoted $0.90/bf for lumber, $1.00/sf for plywood, $1.10/lb for nails, and $0.60 per tie. If the forms will be used three times, the material cost is most nearly:

    (A) $26,136
    (B) $13,349
    (C) $10,645
    (D) $4,258

510. After purchasing a quarry and basic crushing equipment, the contractor is considering an alternative plan to improve the operation of the quarry. The alternative plan will produce an equal amount of crushed rock and equal revenue.

Parameter	Present Plan	Alternative Plan
First Cost ($)	0	10,000
Salvage ($)	0	1,000
Annual Cost ($)	250,000	248,000
Life (years)	—	5

The benefit-cost ratio of the alternative plan (using a 10% rate of return on investment) when compared to the present plan is most nearly:

(A) 0.6
(B) 0.8
(C) 1.0
(D) 1.2

511. A project requires 600 lf of fencing for a storage area. The purchase price of fencing is $4.40/lf. The initial salvage value of a new fence starts at $1.00/lf and is reduced at the rate of 25% per year. A fence can be rented for $1.50/lf per year but has no salvage value. The interest rate is 4%. The project duration (years) at which it first becomes cheaper to purchase rather than to rent is most nearly:

(A) 2.5
(B) 2.8
(C) 3.2
(D) 3.5

512. A restricted area in a refinery prevents access of excavating equipment. A crew of two laborers at \$25/hr each and three laborers at \$35/hr each will be used to excavate 25 yd^3 of earth. If \$4,000 is allowed over a 4-day period for the work, the minimum production rate (yd^3/hr) that is required from the crew is most nearly:

    (A)   1
    (B)   4
    (C)   6
    (D)   26

513. The rigging shown will be used to lift a 60-kip load. The center of gravity of the load is 40 ft from the left end. The force (kips) in Sling A is most nearly:

    (A)   30
    (B)   32.3
    (C)   40
    (D)   43.1

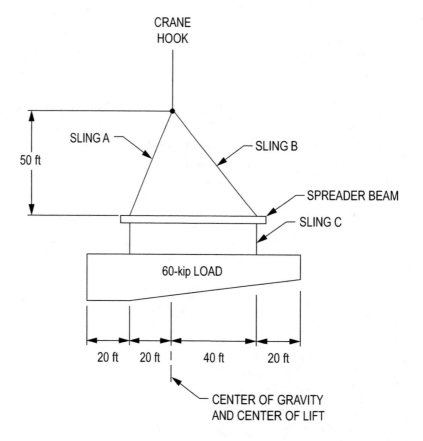

514. The freestanding tower crane shown will be supported on a 16-ft-square concrete foundation. The concrete foundation has a unit weight of 150 pcf. One of the design criteria for the foundation is to prevent overturning of the crane about the toe of the footing. A factor of safety of 1.5 is required on all overturning forces under the following combination of loads:

Pick load, P_p
Boom weight, P_b
Counterweight, P_c
Mast weight, P_m
Wind on boom, W_b
Wind on mast, W_m

Neglecting tower sway, the maximum pick load P_p (kips) is most nearly:

(A) 7.6
(B) 32.1
(C) 49.6
(D) 76.8

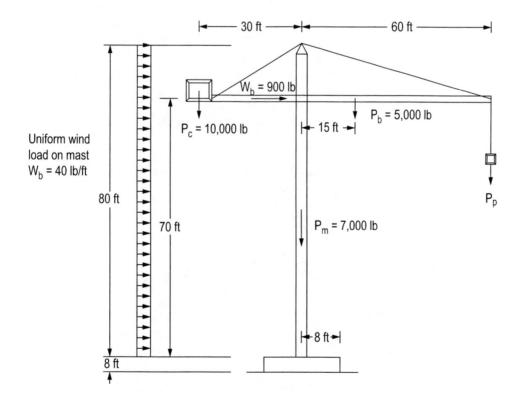

515. Wellpoint systems may be used as a dewatering method under certain conditions. Which of the following statements best describes the conditions where a single-stage wellpoint system is most appropriate?

(A) A deep excavation where the water table must be lowered more than 30 ft but less than 50 ft

(B) Permeable soils where drawdown of the water table is greater than 30 ft but less than 50 ft

(C) A small, shallow excavation in dense or cemented soils

(D) Permeable soils where drawdown of the water table is less than 20 ft

516. An excavation is to be made next to an historic masonry structure as shown in the figure. Which method is least likely to cause damage to the historic structure?

(A) Slurry walls installed deep into the low-permeability soil layer

(B) Soldier piles, beams, and lagging with dewatering wells

(C) Braced sheeting with dewatering wells

(D) Hammer-driven steel sheet piles driven deep into the low-permeability soil layer

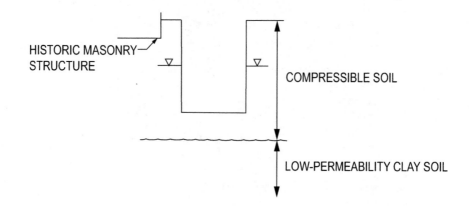

517. A truck hauling earth from a construction site has the following specifications:

Maximum allowable gross vehicle weight	37,800 lb
Empty vehicle weight	10,800 lb
Heaped capacity	12 yd^3
Struck capacity	10 yd^3
Haul time to dump area, including load, haul, return, and dump times	17 min
Delay time	5 min/hr

The soil has the following characteristics:

Bank density	110 pcf
Loose density	100 pcf

The productivity, in bank measure (yd^3/hr), of this operation is most nearly:

(A) 38.8
(B) 35.3
(C) 32.3
(D) 29.5

518. An end-bearing-on-rock pile foundation will be constructed to support a new bridge. The bottom of the pile cap will be at Elevation 980. Pile splices will not be allowed. The pile embedment in the cap is 1'-0". A maximum of a 2-ft cut-off due to driving damage is anticipated. Based on the subsurface exploration log shown, the minimum pile order length (ft) is most nearly:

(A) 50
(B) 55
(C) 60
(D) 65

518. (Continued)

SUBSURFACE EXPLORATION LOG (Sheet 2 of 2)

REGION 3
COUNTY ORANGE
PROJECT INTERSTATE 0
DATE START 5/8/07
DATE FINISH 5/9/07
CASING O.D. 2-1/2" I.D.
SAMPLER O.D. 2" I.D. 1-1/2"
RIG TYPE ACKER B-40
CORE BARREL DOUBLE TUBE

HOLE BAF-3
LINE BASELINE
STA. 93+27
OFFSET 50' RT.
SURF. ELEV. 990.0

HAMMER FALL-CASING 18"
HAMMER FALL-SAMPLER 30"
WEIGHT OF HAMMER-CASING 300 LBS.
WEIGHT OF HAMMER-SAMPLER 140 LBS.

TIME	4 pm	8 am	2 pm
DATE	5/8/07	5/9/07	5/16/07
DEPTH TO WATER	6'	6'	6'

DEPTH BELOW SURFACE	BLOWS ON CASING	SAMPLE NO.	BLOWS ON SAMPLER (0 / .5 / 1.0 / 1.5 / 2.0)	DESCRIPTION OF SOIL AND ROCK	MOIST. CONT. %
35	71 / 79	J9	2 / 2 / 3 / 2	GR - SILTY CLAY	36
	86 / 83				
40	85 / 81	J10	3 / 4 / 3	MOIST - PLASTIC	35
	93 / 91				
45	96				
	121 / 450	J11	20 / 21 / 35	GR - SILTY GRAVEL	10
	391 / 220			MOIST - NON PLASTIC	
50	230 / 200	J12	15 / 36 / 40	52' TO 53' CORED BOULDER RECOVERY 3' MANY FRAGMENTS	5
	370 / 400				
	410 / 380				
60		J13	40 / 60 / 80		7
	REFUSAL	J14	100 REFUSAL @ 60.5'	TOP OF ROCK 60.5' HARD UNWEATHERED BASALT RUN 1 60.5 TO 65.5'-60" - RECOVERY 50'. 12 PIECES RQD 70%	
65				RUN 2 65.5 TO 70.5'-60" - RECOVERY 60'. 6 PIECES RQD 95% HARD UNWEATHERED BASALT	
70				END OF BORING 70.5'	

THE SUBSURFACE INFORMATION SHOWN HEREON WAS OBTAINED FOR STATE DESIGN AND ESTIMATE PURPOSES. IT IS MADE AVAILABLE TO AUTHORIZED USERS ONLY THAT THEY MAY HAVE ACCESS TO THE SAME INFORMATION AVAILABLE TO THE STATE. IT IS PRESENTED IN GOOD FAITH, BUT IT IS NOT INTENDED AS A SUBSTITUTE FOR INVESTIGATIONS, INTERPRETATION OR JUDGMENT OF SUCH AUTHORIZED USERS.

CONTRACTOR SM

DRILL RIG OPERATOR KLINEDINST
SOIL & ROCK DESCRIP. CHASSIE
REGIONAL SOILS ENGR. CHENEY
SHEET 2 OF 2
STRUCTURE NAME/NO. APPLE FREEWAY #2

30   HOLE BAF-3

SUBSURFACE EXPLORATION LOG (Sheet 1 of 2)

REGION 3
COUNTY ORANGE
PROJECT INTERSTATE 0
DATE START 5/8/07
DATE FINISH 5/9/07
CASING O.D. 2-1/2" I.D.
SAMPLER O.D. 2" I.D. 1-1/2"
RIG TYPE ACKER B-40
CORE BARREL DOUBLE TUBE

HOLE BAF-3
LINE BASELINE
STA. 93+27
OFFSET 50' RT.
SURF. ELEV. 990.0

HAMMER FALL-CASING 18"
HAMMER FALL-SAMPLER 30"
WEIGHT OF HAMMER-CASING 300 LBS.
WEIGHT OF HAMMER-SAMPLER 140 LBS.

TIME	4 pm	8 am	2 pm
DATE	5/8/07	5/9/07	5/16/07
DEPTH TO WATER	6'	6'	6'

DEPTH BELOW SURFACE	BLOWS ON CASING	SAMPLE NO.	BLOWS ON SAMPLER (0 / .5 / 1.0 / 1.5 / 2.0)	DESCRIPTION OF SOIL AND ROCK	MOIST. CONT. %
0	0 / 2	J1	1 / 0 / 1	BLACK MUCK WET - PLASTIC	115
	11 / 25	J2	3 / 5 / 7		20
	31 / 40			GR SAND W/ ROOTS AND FIBERS	
	41 / 56	J3	8 / 8 / 9	MOIST - NON PLASTIC	8
10	71 / 83	J4	6 / 5 / 5		29
	91 / 93			GR-BR CLAYEY SILT	
	82 / 81	J5	2 / 3 / 6	MOIST PLASTIC	31
	80 / 87				
	85 / 90				
20	82 / 86	J6	4 / 3 / 3	GR SILTY CLAY	34
	87 / 85			MOIST - PLASTIC	
	90 / 72	J7	2 / 2 / 3		39
	83 / 71				
30	61 / 81	J8	2 / 2 / 2		40
	83 / 72				
	76 / 83				

THE SUBSURFACE INFORMATION SHOWN HEREON WAS OBTAINED FOR STATE DESIGN AND ESTIMATE PURPOSES. IT IS MADE AVAILABLE TO AUTHORIZED USERS ONLY THAT THEY MAY HAVE ACCESS TO THE SAME INFORMATION AVAILABLE TO THE STATE. IT IS PRESENTED IN GOOD FAITH, BUT IT IS NOT INTENDED AS A SUBSTITUTE FOR INVESTIGATIONS, INTERPRETATION OR JUDGMENT OF SUCH AUTHORIZED USERS.

CONTRACTOR SM

DRILL RIG OPERATOR KLINEDINST
SOIL & ROCK DESCRIP. CHASSIE
REGIONAL SOILS ENGR. CHENEY
SHEET 1 OF 2
STRUCTURE NAME/NO. APPLE FREEWAY #2

29   HOLE BAF-3

After Cheney, R.S., and R.G. Chassie, *Soils and Foundations Workshop Manual*, National Highway Institute Course No. 13212, U.S. Department of Transportation, Federal Highway Administration, p. 224, November 1982.

519. The crane shown in the figure has a 180-ft boom and must place two packages on the roof. The goal is to place each package as far from the edge as is safely possible. The following data apply:

> The centerline of rotation of the crane is fixed at 60 ft from the edge of a building.
> The building is 90 ft high.
> Package 1 weighs 17,500 lb.
> Package 2 weighs 39,200 lb.
> The rigging needed for each lift weighs 8,000 lb.
> Interpolation is not allowed for the load table. If a load falls between two entries, choose the more conservative entry.
> Boom standoff from building edge is to be about 7 ft from boom centerline to allow for boom width and safe clearance.

The maximum safe placing distance from the edge of the roof to the centerline of each package is most nearly:

(A)   70 ft for Package 1 and 20 ft for Package 2

(B)   45 ft for Package 1 and 10 ft for Package 2

(C)   51.17 ft for Package 1 and 10.44 ft for Package 2

(D)   30 ft for Package 1 and 10 ft for Package 2

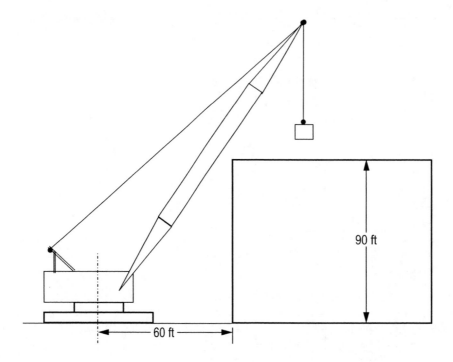

519. (Continued)

Lifting capacities in pounds for a 200-ton, nominal rating, crawler crane with 180 ft of boom*

Radius (ft)	Capacity (lb)	Radius (ft)	Capacity (lb)	Radius (ft)	Capacity (lb)
32	146,300	80	39,200	130	17,900
36	122,900	85	35,800	135	16,700
40	105,500	90	32,800	140	15,500
45	89,200	95	30,200	145	14,500
50	76,900	100	27,900	150	13,600
55	67,200	105	25,800	155	12,700
60	59,400	110	23,900	160	11,800
65	53,000	115	22,200	165	11,100
70	47,600	120	20,600	170	10,300
75	43,100	125	19,200	175	9,600

*Specified capacities based on 75% of tipping loads.

Source: Manitowoc Engineering Co.

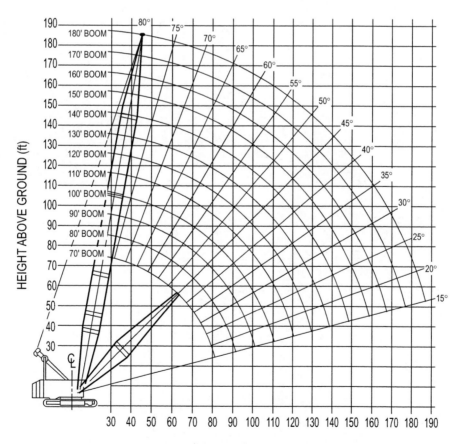

DISTANCE FROM CENTER OF ROTATION (ft)

WORKING RANGES FOR A 200-TON CRAWLER CRANE, NOMINAL RATING.
SOURCE: MANITOWOC ENGINEERING CO.

520. An activity and relationship list for construction of a garage is given below.

Number	Activity	Successors
1	Start	2
2	Sitework	3
3	Footer	4, 10
4	Framing	5, 6, 8, 9
5	Roof	7
6	Sheathing	7, 11
7	Electrical	12
8	Plumbing	10
9	Doors/Windows	12
10	Slab	12
11	Exterior finish	13
12	Interior finish	13
13	Finish	

Which network diagram best represents this project?

(A)

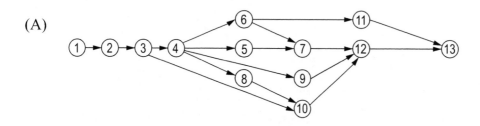

(B)

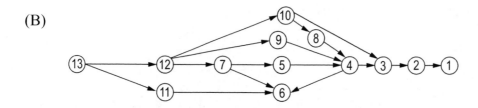

(C)

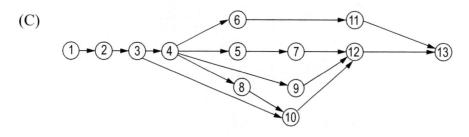

(D)

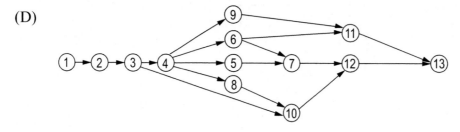

48

521. A workforce of six electricians has to install 420 light fixtures in a new office building. Normally, two electricians working together can install one fixture in 20 min. The crews expect to have a 0.8 efficiency factor due to environmental conditions on this job. The crews will be scheduled for 9-hr workdays, which includes a 1-hr break for lunch. The duration of this construction activity (days, hours) is most nearly:

	Days	Hours
(A)	5	8
(B)	6	5
(C)	7	3
(D)	21	8

522. An activity-on-node network for a project is shown in the figure. All relationships are finish-to-start with no lag unless otherwise noted. If all activities begin at their early start except Activity E, which is delayed by 2 days from its early start, which of the following statements is true?

(A)   Activity E will have no impact on the start time of any other activity.

(B)   Activity E will delay the start of Activity G by 1 day but will not delay project completion.

(C)   Activity E will delay the start of Activity G by 2 days but will not delay project completion.

(D)   Activity E will delay the completion of the project by 2 days.

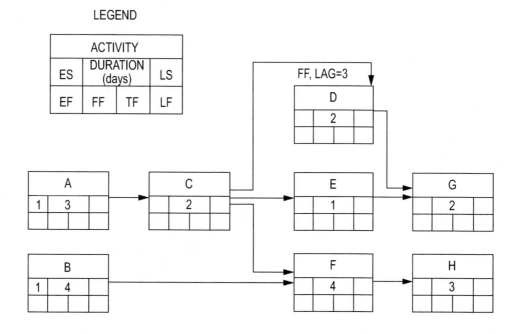

49
GO ON TO THE NEXT PAGE

523. An activity-on-node network for a project is shown. The quantity of a needed crew resource is shown on the activities that require that resource. If the activities cannot be split or interrupted, which histogram represents a resource-leveled schedule for this project?

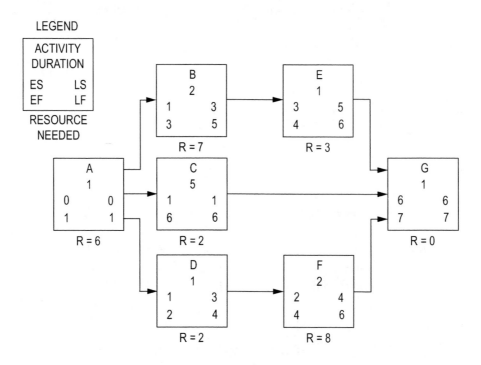

ACTIVITY-ON-NODE NETWORK

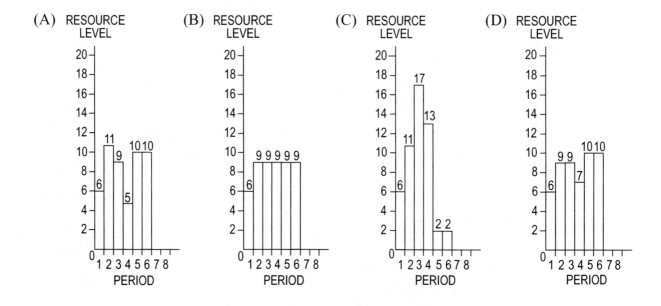

524. A contractor has scheduled work according to the network shown in the figure. Task A is performed by the contractor's employees and will take 4,000 labor hours of effort. The contractor has 10 laborers available to work this task at a standard 40 hr/week and 8 hr/day. Tasks B and C are both subcontracted and they have 4 weeks and 5 weeks, respectively, allocated in their subcontracts. Laborers cost $10/labor hour for straight time and $15/labor hour on overtime. The contractor estimates that it is possible to work 50 hr/week without impacting productivity. General conditions cost $1,000/day every day the contractor or subcontractors are on site. The optimal savings on working overtime is most nearly:

(A) $3,000
(B) $5,000
(C) $6,000
(D) $10,000

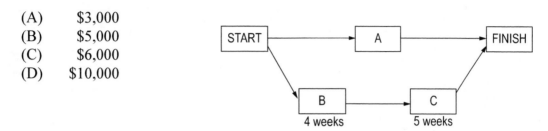

525. Laboratory testing was performed on a soil sample with the following results:

Sieve Analysis Data and Index Properties	
Sieve #	% Passing
3 in.	100
1 1/2 in.	98
3/4 in.	96
#4	77
#10	—
#20	55
#40	—
#100	30
#200	18
Liquid limit	32
Plastic limit	25

According to the Unified Soil Classification System, the classification of the sample is:

(A)   SW

(B)   SP

(C)   SM

(D)   SC

526. A 1/2-in.-diameter, 7-wire prestressing strand is to be stressed in the pretensioning bed shown with a jacking force of 14 tons. The length from the dead-end anchorage to the back of the live-end hydraulic jack is 310 ft. For the strand properties indicated, the expected strand elongation (in.) is most nearly:

(A) 3.5
(B) 12.3
(C) 19.3
(D) 24.7

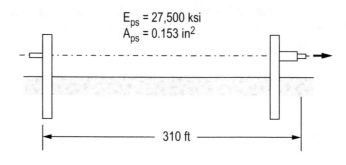

E_{ps} = 27,500 ksi
A_{ps} = 0.153 in^2

310 ft

527. The following weld symbol is indicated on a structural drawing:

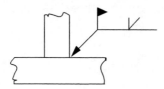

Which option shows the correct weld deposit for this weld symbol?

(A)

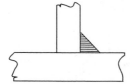

(B)

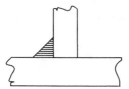

(C)

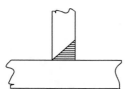

(D)

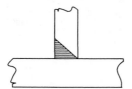

528. Rebar called for in a structural detail states "#8 vertical dowels at 6 in. o.c., alt. #8 to 4 ft AFF with #8 to 7 ft AFF." Which of the following figures shows this detail correctly?

(A)

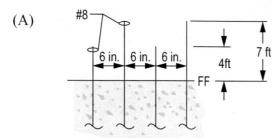

(B)

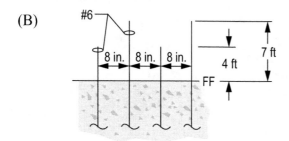

(C)

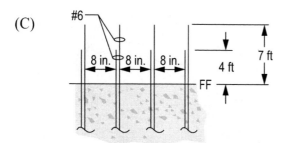

(D)

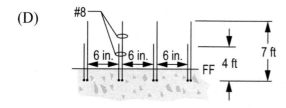

529. A concrete mix design for a highway project requires 590 lb/yd^3 of cement. On a weight basis, the mix design has the proportions 1:2.25:3.25. The aggregates are SSD and have specific gravities of 2.65 for both the fine and the coarse aggregate. The specific gravity of the cement is 3.15. The water/cement ratio is most nearly:

(A) 0.37
(B) 0.46
(C) 0.54
(D) 0.68

530. Immediately following a concrete placement, the concrete temperature is monitored with time. The average temperatures recorded during 4-hour intervals are listed below. If the datum temperature is 20°F, the concrete maturity (°F-hours) at the end of 48 hours is most nearly:

(A)   632
(B)   960
(C)   2,528
(D)   3,360

Time (hours)	Temperature (°F)
0–4	66
4–8	70
8–12	76
12–16	80
16–20	82
20–24	82
24–28	78
28–32	74
32–36	72
36–40	68
40–44	64
44–48	60

531. A 4-ft-tall × 60-ft-long × 20-ft-wide concrete block will be placed at the bottom of the lake. Assume 100% contact with the lake bed and uniform loading. The density of the concrete is 150 lb/ft^3. The resulting contact pressure (lb/ft^2) will be most nearly:

(A)   350
(B)   600
(C)   3,000
(D)   9,000

532. Formwork for a 9-ft-high concrete wall is required. The rate of placement is estimated to be 15 ft/hr with normal internal vibration and vibrator immersion not exceeding 4 ft. The concrete to be used contains Type II cement and a retarder and weighs 155 lb/ft^3. The concrete will be placed at a concrete temperature of 80°F. The maximum design lateral concrete pressure (lb/ft^2) based on ACI 347-04 is most nearly:

(A)   1,302
(B)   1,395
(C)   1,461
(D)   1,561

533. A 5-ft-wide × 70-ft-tall scaffold tower is erected next to an existing structure to allow the installation of a brick veneer. The minimum lateral anchorage shown meeting the requirements of federal OSHA regulation 29 CFR 1926.451, Subpart L, is most nearly:

(A)

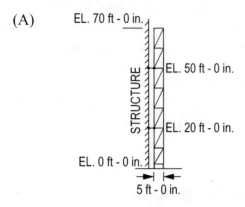

(B)

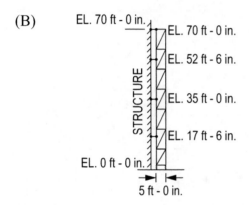

(C)

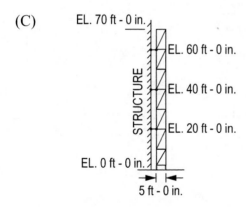

(D)

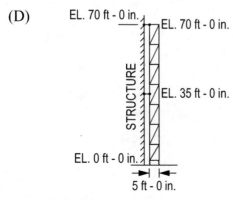

534. A multistory concrete building is being constructed using two levels of shores and forms and two levels of reshores. Construction has been completed through Level 8 as shown in the figure. An analysis has been made to the step shown using the assumption that the shores and reshores are infinitely stiff, the slabs have equal stiffness, and the reshores are initially installed snug tight. Actual (or unfactored) loads and forces are expressed as a multiple of the typical floor dead load, D. The forms and shores combined weigh 0.10 D at each level installed, and the reshores weigh 0.05 D at each level installed. At the end of the previous operation shown, there is no worker live load. In the next operation, the forms and shores between Levels 6 and 7 are stripped, flown upward, and stored on top of Level 8 for re-erection. The load of workers imposed on Levels 6 and 8 during this next operation is 0.40 D on each. The reshores will be moved from between Levels 4 and 5 to between 6 and 7 in a future operation.

As the operation to move the forms and shores from between Levels 6 and 7 to Level 8 is ending, but while the workers are still on the floor at Level 8, the load carried by the floor at Level 7 is most nearly:

(A)   1.40 D
(B)   1.45 D
(C)   1.65 D
(D)   1.70 D

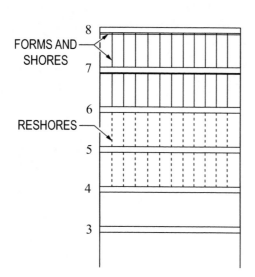

Shore Loads Before Step	Loads Carried by Floors in Multiples of D			Shore Loads After Step
	Before Step	Change During Step	After Step	
	0.00 D			
1.10 D				
	0.70 D			
1.50 D				
	1.90 D			
0.65 D				
	1.25 D			
0.45 D				
	1.45 D			

535. A bracket is anchored to a concrete wall using a bolt screwed into an imbedded insert as shown in the figure. The tension (kips) in the bolt is most nearly:

(A) 0
(B) 4.5
(C) 12
(D) 32

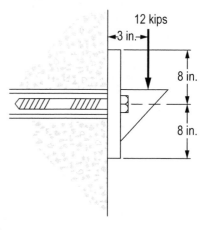

536. A trench is to be excavated in a soil that has been determined to be Type A in accordance with federal OSHA regulation Subpart P. The trench excavation will not exceed 12 ft in depth and will be open for only 8 hours from the start of excavation to completion of backfilling. The maximum slope allowed by OSHA for the trench walls under these conditions is most nearly:

(A) vertical

(B) 0.5 horizontal to 1.0 vertical

(C) 0.75 horizontal to 1.0 vertical

(D) 1.0 horizontal to 1.0 vertical

537. For the formwork shown the live load is 50 lb/ft^2. Joists are single-span simply supported. Stringers are two-span continuous beams (must consider continuity). Neglecting the weight of the formwork, the load (lb) on Shore B2 is most nearly:

(A) 16,875
(B) 13,500
(C) 2,344
(D) 1,875

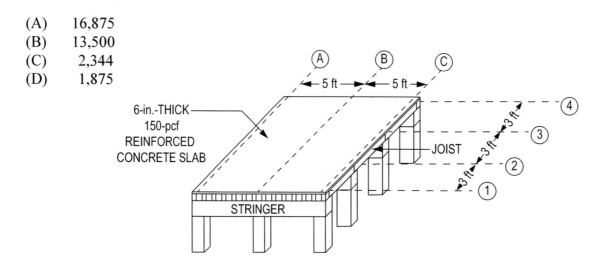

538. An employee is assigned for an entire 8-hour shift to remove rivets by burning off the heads and punching the rivets out with power tools. The rivets are coated with lead paint. This work has been determined to generate an airborne concentration of lead of 1,500 μg/m^3 (μg/m^3 = micrograms per cubic meter). The **minimum** level of respiratory protection to this exposure required by federal OSHA regulation 1926.62, Subpart D, is:

(A) half-mask air-purifying respirator with appropriate filters

(B) helmet supplied-air respirator operated in continuous-flow mode

(C) full-facepiece self-contained breathing apparatus (SCBA) operated in positive-pressure mode

(D) half-mask or full-facepiece supplied-air respirator operated in continuous-flow mode

GO ON TO THE NEXT PAGE

539. A construction company has 750,000 employee hours worked with the following safety record:

Incidence Category	No. of Incidences
Minor injuries (first aid only)	10
Medical-only injuries (no lost time or light duty)	4
Medical injuries resulting in "light duty" restrictions	3
Lost-time injuries	5

The OSHA Incidence Rate for recordable cases is most nearly:

(A)   1.33
(B)   2.13
(C)   3.20
(D)   5.86

540. Temporary traffic control measures are required as shown on the figure. The lane width is 12 ft, and the posted speed limit is 35 mph. According to MUTCD, Part 6, the minimum length (ft) of Zone A is most nearly:

(A) 12.3
(B) 94.8
(C) 245
(D) 420

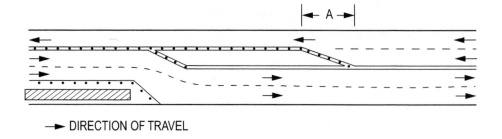

DIRECTION OF TRAVEL

CONSTRUCTION ZONE ON URBAN COLLECTOR

This completes the afternoon session. Solutions begin on page 83.

Answers to the Civil AM Practice Exam

Detailed solutions for each question begin on the next page.

101	D	**121**	A
102	C	**122**	A
103	C	**123**	C
104	D	**124**	B
105	C	**125**	C
106	B	**126**	B
107	C	**127**	C
108	D	**128**	B
109	B	**129**	D
110	B	**130**	D
111	A	**131**	C
112	A	**132**	D
113	C	**133**	A
114	D	**134**	C
115	B	**135**	B
116	D	**136**	C
117	C	**137**	B
118	D	**138**	D
119	A	**139**	C
120	C	**140**	C

CIVIL AM SOLUTIONS

101. Reference: Peurifoy and Oberlender, *Estimating Construction Costs,* 8th ed., Chapter 10, p. 273, Quantity Takeoff.

$$\text{Horizontal length of side slope} = 14 \times \frac{3}{2} = 21.0 \text{ ft}$$

$$\text{Slope length} = \sqrt{(14)^2 + (21)^2} = 25.24 \text{ ft}$$

$$\text{Cross-sectional area of lining} = \left[(2 \times 25.24) + 9\right]\frac{7}{12} = 34.70 \text{ ft}^2$$

$$\text{Volume of lining} = \frac{(34.70 \times 227)}{27} = 291.7 \text{ yd}^3$$

$$\text{Delivered volume} = 291.7 \text{ yd}^3 \times \underset{\text{(waste)}}{1.12} = 327 \text{ yd}^3$$

THE CORRECT ANSWER IS: (D)

102. Reference: Nunnally, *Construction Methods and Management*, 8th ed., 2011, p. 299.

$$D = \frac{\$75,000 - \$10,000}{10}$$

$$D = \$6,500$$

$$\text{Book value after 8 years} = \$75,000 - (8)(\$6,500) = \$23,000$$

THE CORRECT ANSWER IS: (C)

103. Reference: AGC, *Construction Planning and Scheduling*, pub. 3500.1, 6th ed., p. 37.

$$\text{Crew cost} = 2(\$50/\text{hr}) = \$100/\text{hr}$$

$$\text{Days allowed} = \frac{\$4,000}{(8 \text{ hr/day})(\$100/\text{hr})} = 5 \text{ days}$$

THE CORRECT ANSWER IS: (C)

104. Reference: Nunnally, *Construction Methods and Management*, 8th ed., 2011, pp. 282–285.

Activities: ⑦ + ④ + ⑤
Days: 30 + 10 + 10 = 50 days

THE CORRECT ANSWER IS: (D)

105. Reference: Ricketts, Loftin, and Merritt, *Standard Handbook for Civil Engineers*, 5th ed., p. 4.11.

$$1{,}000 \text{ kN} = 1{,}000 \text{ kN} \times \frac{1 \text{ ton}}{8.896444 \text{ kN}} = 112.4 \text{ tons}$$

150 tons > 112.4 tons

THE CORRECT ANSWER IS: (C)

106. Reference: Shapiro, Shapiro, and Shapiro, *Cranes and Derricks*, 3rd ed., 2000, p. 244.

$\tan(x) = \dfrac{40}{30}$ $x = 53.13°$

$\cos(51.13°) * 100 \text{ ft} = 60 \text{ ft}$

$60 \text{ ft} - 35 \text{ ft} = 25 \text{ ft}$

THE CORRECT ANSWER IS: (B)

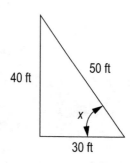

107. Reference: Hurd, *Formwork for Concrete*, ACI SP-4, 7th ed., 2005.

$$w = (20 \text{ lb/ft}^2)(8 \text{ ft}) = 160 \text{ lb/vertical ft per brace location}$$

$$\sum M_a = 0$$

$$\sum M_a = (160 \text{ lb/ft})(16 \text{ ft})(16 \text{ ft/2}) - 10 \text{ ft } (R_x) = 0$$

$$R_x = 2,048 \text{ lb}$$

$$\text{Axial load in brace} = \frac{(2,048)\sqrt{2}}{1} = 2,896 \text{ lb}$$

THE CORRECT ANSWER IS: (C)

108. Reference: NAVFAC, DM 7.2-60.

The wall translation (or strain) required to achieve the passive state is at least twice that required to reach the active state.

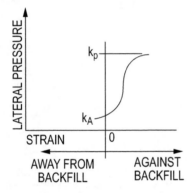

THE CORRECT ANSWER IS: (D)

109. The solution is based on the knowledge that consolidation settlement is the result of the expulsion of pore water from saturated soil due to imposed load. Therefore, the volume of the wick drain effluent (water) to be treated equals the consolidation settlement volume over the affected site area, and is computed as follows:

Affected area $= 21.5 \text{ acres} \times 43,560 \text{ ft}^2/\text{acre} = 936,540 \text{ ft}^2$

Mean consolidation settlement over affected area $= 22 \text{ in.} = 1.83 \text{ ft}$

Settlement volume = effluent volume $= 936,540 \text{ ft}^2 \times 1.83 = 1,713,868 \text{ ft}^3$

Convert to gal: $1,713,868 \text{ ft}^3 \times 7.48 \text{ gal/ft}^3$ $= 12,819,733 \text{ gal}$

Cost for effluent treatment and disposal $= 12,819,733 \text{ gal} \times \$0.25/\text{gal}$

$= \$3,204,934$

THE CORRECT ANSWER IS: (B)

CIVIL AM SOLUTIONS

110. Reference: Terzaghi, Peck, Mesri, *Soil Mechanics in Engineering Practice*, 3rd ed., p. 84,

Effective vertical stress at Point A, σ'_v

$$= 10\,\text{ft} \times 120\,\text{pcf} + 5\,\text{ft}\left(120\,\text{pcf} - 62.4\,\text{pcf}\right) + 7\,\text{ft}\left(110\,\text{pcf} - 62.4\,\text{pcf}\right)$$
$$= 1,200\,\text{psf} + 288\,\text{psf} + 333\,\text{psf}$$
$$= 1,821\,\text{psf}$$

THE CORRECT ANSWER IS: (B)

111. The ultimate bearing capacity would be based on buoyant unit weight, also referred to as the effective unit weight.

Effective unit weight = saturated unit weight – unit weight of water

THE CORRECT ANSWER IS: (A)

112. References: Coduto, *Foundation Design Principles and Practice*, 2nd ed., p. 250.

The long-term settlement for Case I is less than Case II because clay is subject to long-term settlement.

THE CORRECT ANSWER IS: (A)

113. References: Day, *Geotechnical and Foundation Engineering*, 1999, p. 10-27, and NAVFAC 7.1-329.

The minimum factor of safety for permanent slopes is 1.5. Other references use a factor of safety greater than or equal to 1.3, but of the options presented 1.5 is the closest.

THE CORRECT ANSWER IS: (C)

CIVIL AM SOLUTIONS

114. Since the structure is cantilevered, in addition to the wind, dead load and live load will contribute to uplift.

THE CORRECT ANSWER IS: (D)

115. By inspection, Member b = 0 kips, and Member c = 100 kips.

THE CORRECT ANSWER IS: (B)

116. Beam stress, $f = M/S$, where $M = wL^2/8$ and $S = bh^2/6$.
S is equal for both beams, but M varies because it depends on beam length.

Beam 1 (shorter beam): $M_1 = wL^2/8$
Beam 2 (longer beam): $M_2 = w(2L)^2/8 = 4wL^2/8$

M_2 is four times greater than M_1. Therefore the maximum bending stress is four times greater in the longer beam.

THE CORRECT ANSWER IS: (D)

117. Uniform load: $V = \dfrac{wL}{2} = \dfrac{1(30)}{2} = \dfrac{30 \text{ kips}}{2} = 15 \text{ kips}$

Point load: $V = \dfrac{P}{2} = 15 \text{ kips}$
$P = 2(15) = 30 \text{ kips}$

THE CORRECT ANSWER IS: (C)

118. I_x is maximum for this section by inspection, or calculate $I_x \approx \Sigma A d^2$ for each section.

THE CORRECT ANSWER IS: (D)

119. $\phi = 32°$ $\qquad K_a = \tan^2(45 - \phi/2) = 0.307$

$\gamma_t = 110 \text{ pcf}$ $\quad P_a = (0.5)(110)(8)^2(0.307) = 1,081 \text{ lb/ft}$

$\qquad\qquad M_a = (1,081)(8/3) = 2,883 \text{ ft-lb/ft}$

$\left. \begin{array}{l} (2)(8)(150)(1)(3) = 7,200 \text{ ft-lb/ft} \\ (1/2)(2)(8)(150)(1)(2)(2/3) = 1,600 \text{ ft-lb/ft} \end{array} \right\}$ $\text{total} = 8,800 \text{ ft-lb/ft}$

$SF = 8,800/2,883 = 3.05$

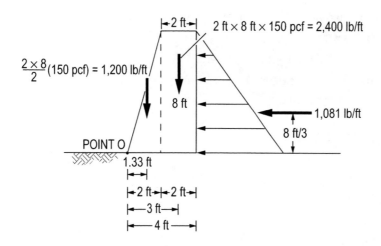

THE CORRECT ANSWER IS: (A)

120. Reference: Mott, *Applied Fluid Mechanics*, 6th ed., 2005, p. 450.

$$Q = VA = \left\{ \frac{1.49}{n} R^{2/3} S^{1/2} \right\} A$$

$$= \left\{ \frac{1.49}{0.022} \left[\frac{(1.5 \text{ ft} \times 4 \text{ ft})}{4 \text{ ft} + 2(1.5 \text{ ft})} \right]^{2/3} (0.002)^{1/2} \right\} (1.5 \text{ ft} \times 4 \text{ ft})$$

$$= 16.4 \text{ cfs}$$

$$\text{Volume} = 25 \text{ acre-ft} \times \frac{43,560 \text{ ft}^3}{1 \text{ acre-ft}} = 1.089 \times 10^6 \text{ ft}^3$$

$$\text{Time} = \frac{1.089 \times 10^6 \text{ ft}^3}{16.4 \text{ ft}^3/\text{sec}} \times \frac{1 \text{ min}}{60 \text{ sec}} \times \frac{1 \text{ hr}}{60 \text{ min}}$$

$$= 18.5 \text{ hours}$$

THE CORRECT ANSWER IS: (C)

121.

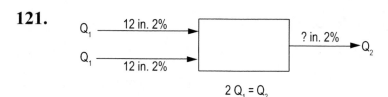

$$2Q_1 = Q_2$$

Reference: Viessman and Lewis, *Introduction to Hydrology*, 4th ed., 1996, p. 252.

$$2[V_1 A_1] = [V_2 A_2]$$

$$2\left[\left(\frac{1.49}{n}\right)(A_1) R_1^{2/3} S^{1/2}\right] = \left[\left(\frac{1.49}{n}\right)(A_2) R_2^{2/3} S^{1/2}\right]$$

$$2\left[(A_1)\left(\frac{A_1}{P_1}\right)^{2/3}\right] = \left[(A_2)\left(\frac{A_2}{P_2}\right)^{2/3}\right]$$

$$A_1 = \frac{\pi D^2}{4} = \frac{\pi(1)^2}{4} = 0.785 \text{ ft}^2$$

$$P_1 = \pi(D) = \pi(1) = 3.14 \text{ ft}$$

$$2\left[(0.785)\left(\frac{0.785}{3.14}\right)^{2/3}\right] = \left[\left(\frac{\pi D_2^2}{4}\right)\left(\frac{\frac{\pi(D_2)^2}{4}}{(\pi D_2)}\right)^{2/3}\right]$$

$$0.623 \qquad = \left(\frac{\pi D_2^2}{4}\right)\left(\frac{D_2}{4}\right)^{2/3}$$

$$= \pi\left(\frac{D_2^2}{4}\right)\left(\frac{D_2}{4}\right)^{2/3}$$

$$= \pi(D_2)^{8/3}\left(\frac{1}{4}\right)\left(\frac{1}{4}\right)^{2/3}$$

$$0.623 = 0.311(D_2)^{8/3}$$

$$\left(\frac{0.623}{0.311}\right)^{3/8} = D_2$$

$$D_2 = 1.297 \text{ ft} \times \frac{12 \text{ in.}}{\text{ft}} = 15.6 \text{ in.} \approx 16 \text{ in.}$$

THE CORRECT ANSWER IS: (A)

CIVIL AM SOLUTIONS

122.   Reference: Mays, *Water Resources Engineering*, 2001, p. 211.

According to the arithmetic mean method, the average precipitation is simply the average of all the rainfall gages.

Average precipitation = (2.1 + 3.6 + 1.3 + 1.5 + 2.6 + 6.1 + 5.1 + 4.8 + 4.1 + 2.8 + 3.0)/11
Average precipitation = 3.4 in.

THE CORRECT ANSWER IS: (A)

123.   Reference: *Water Supply and Pollution Control,* Viessman and Hammer, 6th ed., 1998, p. 229.

From the IDF curve, read a rainfall intensity of 3.5 in./hr for a 50-year frequency rainfall with a 60-min duration.

From the table, the runoff coefficient for a downtown area is 0.70 – 0.95. For the maximum runoff rate, use the high value of 0.95.

$Q = CiA = 0.95 \times 3.5$ in./hr $\times 90$ ac

$Q = 300$ cfs

THE CORRECT ANSWER IS: (C)

124.   Reference: Davis and Cornwell, *Introduction to Environmental Engineering*, 4th ed., 2008, p. 61.

$$\text{Time} = \frac{V}{Q}$$

$$V = 400,000 \text{ gal} \times \frac{\text{ft}^3}{7.48 \text{ gal}} = 53,476 \text{ ft}^3$$

$$Q = 1.5 \text{ ft}^3/\text{sec}$$

$$\text{Time} = \frac{53,476 \text{ ft}^3}{1.5 \text{ ft}^3/\text{sec}} \times \frac{1 \text{ hr}}{3,600 \text{ sec}} = 9.9 \text{ hours}$$

THE CORRECT ANSWER IS: (B)

125. Reference: Merritt, Loftin, and Ricketts, *Standard Handbook for Civil Engineers*, 4th ed., 1996, pp. 21.22 and 21.42.

The Darcy-Weisbach equation is $h_f = f \dfrac{L}{D} \dfrac{V^2}{2g}$

where

h_f = headloss, ft

f = friction factor, unitless

L = length, ft

D = diameter of pipe, ft

V = velocity, ft/sec

g = gravitational constant, 32.2 ft/sec^2

Substituting gives

$$5 \text{ ft} = 0.0115 \times \frac{1{,}650 \text{ ft}}{3.0 \text{ ft}} \times \frac{V^2}{2 \times 32.2 \text{ ft/sec}^2}$$

$$V^2 = 50.91 \text{ ft}^2/\text{sec}^2$$

$$V = 7.135 \text{ ft/sec}$$

$$Q = VA = V \times \frac{\pi}{4} D^2 = 7.135 \text{ ft/sec} \times \frac{\pi}{4}(3.0 \text{ ft})^2$$

$$Q = 50 \text{ cfs}$$

THE CORRECT ANSWER IS: (C)

126. Reference: Lin, Shundar, and C.C. Lee, *Water and Wastewater Calculations Manual*, 2001, p. 240.

$$z_1 + \frac{P_1}{\gamma} + \frac{v_1^2}{2g} = z_2 + \frac{P_2}{\gamma} + \frac{v_2^2}{2g}$$

$z_1 = z_2$

Since $A_1 > A_2$, $v_1 < v_2$.

$$\therefore \frac{v_1^2}{2g} < \frac{v_2^2}{2g}$$

so $P_1 > P_2$ to balance

THE CORRECT ANSWER IS: (B)

127. Reference: Hickerson, *Route Location and Design*, 5th ed., p. 64.

$R = 5{,}729.648/D_C^\circ$

$ = 5{,}729.648/10 = 572.96 \text{ ft}$

$T = R \tan\left(\frac{1}{2}\Delta\right) = R \tan(6.25^\circ)$

$ = 572.96 \,(\tan 6.25^\circ)$

$ = 572.96 \,(0.1095178)$

$ = 62.75 \text{ ft}$

Station PC $= $ Station PI $- T$

$\phantom{\text{Station PC}} = [12+40] - 62.75$

$\phantom{\text{Station PC}} = 11+77.25$

Station PT $=$ Station PC $+$ length of curve

Length of curve $= L = 100\,\Delta/D_C^\circ$

$\phantom{\text{Length of curve}} = 100(12.5)/10 = 125 \text{ ft}$

Station PT $=$ Station PC $+ 125 \text{ ft} = [11+77.25] + 125 = 13+02.25$

PI 12+40

T

DEFLECTION = 12° 30′ = 12.5°

PC

PT

$D_C^\circ = 10^\circ$

NOT TO SCALE

THE CORRECT ANSWER IS: (C)

128. Reference: Hickerson, *Route Location and Design*, 5th ed., pp. 154, 160.

$L = KA$

$K = L/A$

L = length of vertical curve, ft

A = algebraic difference in grades, percent ($g_2 - g_1$)

Given: VPC = 12+00

VPI = 13+50

VPT = 15+00

$g_1 = -2.30\%$

$g_2 = +3.00\%$

L = 300 ft

$K = \dfrac{L}{A} = \dfrac{300}{3 - (-2.3)} = 56.60$ ft/percent for the vertical curve.

The length from Station 14+00 to Station 15+00 = 100 ft

$K = \dfrac{L}{A}$

$A = \dfrac{L}{K} = \dfrac{100}{56.60} = 1.77\%$

$A = g_2 - g_1$

Tangent slope at Station 14+00 = g_1

$g_1 = g_2 - A = 3.00\% - 1.77\% = 1.23\%$

Alternate solution:

Y = elevation at a point X ft from VPC

Y′ = slope at a point X ft from VPC

$X = [14 + 00] - [12 + 00] = 200$ ft

g_1 = slope 1 in ft/ft

g_2 = slope 2 in ft/ft

L = length of vertical curve, ft

$Y = Y_{VPC} + g_1 X + \left(\dfrac{g_2 - g_1}{2L} \right) X^2$

$Y' = g_1 + \left(\dfrac{g_2 - g_1}{L} \right) X$

$Y' = -0.023 + \left(\dfrac{0.03 - (-0.023)}{300} \right) 200 = 0.0123$ ft/ft or 1.23%

THE CORRECT ANSWER IS: (B)

129. Reference: Garber and Hoel, *Traffic & Highway Engineering,* 4th ed., pp. 130–132.

$$\text{AADT} = \frac{\sum \left(\text{Jan. through Dec.} \right)}{12}$$
$$= 833,200 \, / \, 12 = 69,433$$

$\sum \left(\text{June through Aug.} \right) = 77,300$

$$78,950$$
$$\underline{77,200}$$
$$233,450 \, / \, 3 = 77,817$$

Seasonal factor for June through August

$$= 77,817 \, / \, 69,433$$
$$= 1.121$$

THE CORRECT ANSWER IS: (D)

130. Reference: Garber and Hoel, *Traffic & Highway Engineering*, 3rd ed., p. 841.

The commonly used soil classification systems for engineering applications are USCS and AASHTO. Both of these systems use gradation and Atterberg limits as two of the criteria.

THE CORRECT ANSWER IS: (D)

131. Reference: Coduto, Yeung, and Kitch, *Geotechnical Engineering: Principles and Practices,* 2nd ed., p. 184.

The Standard Penetration Test (SPT) N-value provides an indication of the relative density of cohesionless soils.

THE CORRECT ANSWER IS: (C)

132. Reference: *Design and Control of Concrete Mixtures*, 14th ed., p. 242.

An early-strength concrete is needed with a minimum compressive strength of 3,500 psi. To achieve the requirements, a Type III cement and chemical accelerators would be necessary.

THE CORRECT ANSWER IS: (D)

133. Reference: NCEES, *FE Reference Handbook,* 9.2.

Reduction in strength due to cyclical loads

THE CORRECT ANSWER IS: (A)

134. Area = $\pi d^2/4 = 28$ in^2

Compressive stress = axial load/area

Sample 1 $f'_c = \dfrac{65,447}{28} = 2,313$ psi

Sample 2 $f'_c = \dfrac{63,617}{28} = 2,248$ psi

Sample 3 $f'_c = \dfrac{79,168}{28} = 2,797$ psi

Average $= \dfrac{(2,313 + 2,248 + 2,797)}{3} = 2,452$ psi

THE CORRECT ANSWER IS: (C)

135. Reference: Garber and Hoel, *Traffic & Highway Engineering,* 4th ed., p. 901.

$$\text{Total density}\,(\gamma) = \frac{W}{V} = \frac{W_s + W_w}{V_s + V_w + V_a}$$

where γ = total density

W = total weight

V = total volume

W_s = weight soil

W_w = weight of water

V_s = volume of soil

V_w = volume of water

V_a = volume of air

$$\gamma = \frac{9.11\,\text{lb} - 4.41\,\text{lb}}{0.03\,\text{ft}^3} = 156.67\,\text{lb/ft}^3\,(\text{pcf})$$

$$\text{Dry unit weight of soil}\,(\gamma_d) = \frac{\gamma}{1 + w}$$

where w = moisture content

$$\gamma_d = \frac{156.67\,\text{pcf}}{1 + 0.115} = 140.51\,\text{pcf}$$

THE CORRECT ANSWER IS: (B)

136. Reference: Kavanagh, *Surveying with Construction Applications*, 6th ed., 2007, pp. 569–573.

Use Average End Area Method.

Stationing	Excavation (yd³)	Embankment (yd³)
1+00 to 2+00	$\dfrac{50+150}{2} \times \dfrac{100}{27} = 370$	
2+00 to 3+00	$\dfrac{50+0}{2} \times \dfrac{100}{27} = 93$	$\dfrac{0+40}{2} \times \dfrac{100}{27} = 74$
Total	**463**	**74**

Net excess excavated material $= 463 - 74 = 389$ yd³

THE CORRECT ANSWER IS: (C)

137. Reference: Kavanagh, *Surveying with Construction Applications*, 6th ed., 2007, pp. 493–501.

Existing:

$\Delta H = (2 + 88.4) - (0 + 23.0) = 288.4 - 23.0 = 265.4$ ft

$\Delta V = 630.32 - 609.39 = 20.93$ ft

New:

$\Delta H = (1 + 15.0) - (0 + 23.0) = 115.0 - 23.0 = 92$ ft

$\Delta V = \dfrac{92}{265.4} \times 20.93 = 7.26$ ft

Inv Elev. $= 630.32 - 7.26 = 623.06$ ft

The top of the pipe will be above the invert elevation by (60 in. – 6 in.)/12 in./ft = 4.50 ft

$623.06 + 4.50 = 627.56$ ft

THE CORRECT ANSWER IS: (B)

138. Reference: *Developing Your Stormwater Pollution Prevention Plan*, USEPA, May 2007, p. 3. Victor Miguel Ponce, *Engineering Hydrology*, 1st ed., p. 538.

Rushing erosion is not identified in either reference.

THE CORRECT ANSWER IS: (D)

139. Reference: OSHA 29 CFR 1926, Subpart P, Appendix B.

Type B soil has a maximum permissible slope of 1:1.

Therefore, a 12-ft depth requires a 12-ft distance.

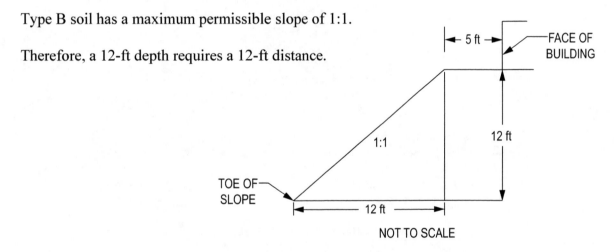

Since there is a 5-ft perimeter strip, the minimum distance from the toe of the slope to the face of the structure = 12 ft +5 ft = 17 ft.

THE CORRECT ANSWER IS: (C)

CIVIL AM SOLUTIONS

140. Reference: AASHTO: *Roadside Design Guide,* 4th ed., 2011, pp. 3-9 and 3-10.

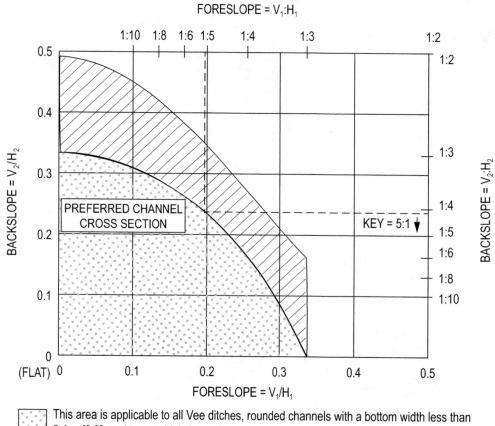

This area is applicable to all Vee ditches, rounded channels with a bottom width less than 2.4 m [8 ft], and trapezoidal channels with bottom widths less than 1.2 m [4 ft].

This area is applicable to rounded channels with bottom width of 2.4 m [8 ft] or more and to trapezoidal channels with bottom widths equal to or greater than 1.2 m [4 ft].

Adapted from AASHTO *Roadside Design Guide,* 4th edition, 2011.

THE CORRECT ANSWER IS: (C)

Answers to the CONSTRUCTION PM Practice Exam

Detailed solutions for each question begin on the next page.

501	B	521	C
502	B	522	A
503	A	523	D
504	D	524	A
505	B	525	C
506	A	526	D
507	D	527	D
508	D	528	A
509	C	529	B
510	B	530	C
511	B	531	A
512	A	532	B
513	D	533	C
514	B	534	C
515	D	535	B
516	A	536	B
517	D	537	C
518	B	538	D
519	D	539	C
520	A	540	C

CONSTRUCTION PM SOLUTIONS

501. References: Peurifoy, Schexnayder, and Shapira, *Construction Planning, Equipment and Methods,* 7th ed., Chapters 4 and 10; and Nunnally, *Construction Methods and Management,* 8th ed., Chapters 2, 3, and 4.

Density of embankment fill, $\gamma_{dry} = (0.90)(120.0 \text{ pcf}) = 108.0 \text{ pcf}$

Total weight of dry soil required:

$$W_{total} = (500,000 \text{ yd}^3)(27 \text{ ft}^3/\text{yd}^3)(108.0 \text{ pcf}) = 1.458 \times 10^9 \text{ lb}$$

Dry unit weight of soil in the truck:

$$\gamma_{dry} = G_s \gamma_w / (1 + e) = (2.65)(62.4 \text{ pcf})/(1 + 1.30) = 71.9 \text{ pcf}$$

Truck capacity:

$$W_{truck} = (5.0 \text{ yd}^3)(27 \text{ ft}^3/\text{yd}^3)(71.9 \text{ pcf}) = 9,700 \text{ lb/truck}$$

Therefore, the minimum number of trucks required is:

$$N = W_{total}/W_{truck} = 1.458 \times 10^9 / 9,700 = 150,000 \text{ trucks}$$

THE CORRECT ANSWER IS: (B)

502. Reference: Anderson and Mikhail, *Surveying–Theory and Practice,* 7th ed., pp. 937–940.

$$
\begin{aligned}
V &= \left(\frac{1.0 + 0.5 + 1.5 + 1.5}{4} + \frac{0.5 + 0.0 + 1.0 + 1.5}{4} + \frac{1.5 + 1.5 + 2.0 + 2.0}{4} + \frac{1.5 + 1.0 + 1.5 + 2.0}{4} \right. \\
&\quad + \frac{1.0 + 0.0 + 0.3 + 1.5}{4} + \frac{2.0 + 2.0 + 2.5 + 3.0}{4} + \frac{2.0 + 1.5 + 1.5 + 2.5}{4} \\
&\quad \left. + \frac{1.5 + 0.3 + 0.7 + 1.5}{4} \right)(10 \times 15) \\
&= \left(\frac{4.5 + 3.0 + 7.0 + 6.0 + 2.8 + 9.5 + 7.5 + 4.0}{4} \right)(10 \times 15) \\
&= \frac{44.3}{4}(150) \\
&= 1,661 \text{ ft}^3 \\
&= 61.5 \text{ yd}^3
\end{aligned}
$$

THE CORRECT ANSWER IS: (B)

CONSTRUCTION PM SOLUTIONS

503. Reference: Kavanagh, *Surveying with Construction Application*, 6th ed., Sections 11.2 and 11.3, pp. 384–386.

$$M = R\left(1 - \cos\frac{\Delta}{2}\right)$$

$$12.8\text{ ft} = 200\text{ ft}\left(1 - \cos\frac{\Delta}{2}\right)$$

$$\Delta = 41.22°$$

$$L = 2\pi R\frac{\Delta}{360°}$$

$$= 2\pi(200\text{ ft})\left(\frac{41.22°}{360°}\right)$$

$$= 143.88\text{ ft}$$

THE CORRECT ANSWER IS: (A)

504. Reference: Schexnayder and Mayo, *Construction Management Fundamentals,* 2004, Chapter 6.

B is the turning point where the job goes from excavation to a fill operation. D is the point where it goes back to excavation. Therefore, B–D is the fill operation (Statement II), and D is a transition point (Statement III). Statements II and III are true.

THE CORRECT ANSWER IS: (D)

505. Reference: Peurifoy, Schexnayder, and Shapira, *Construction Planning, Equipment and Methods*, 7th ed., p. 98.

Both shrink and swell can be estimated. Shrink is the relationship between the in-place (bank) and compacted states. Dry density in-place can be estimated from tests on rings. Compacted dry density can be estimated from ASTM D698. Swell can be estimated from gradation (soil type).

THE CORRECT ANSWER IS: (B)

506. Reference: Peurifoy, Schexnayder, and Shapira, *Construction Planning, Equipment and Methods*, 7th ed., p. 80.

Total amount of material is sum of absolute values of peaks and low points.

$$= (5,000 - 2,000) + (3,000 - 2,000) + (2,000 - 0)$$
$$= 3,000 + 1,000 + 2,000 = 6,000 \text{ bcy}$$

THE CORRECT ANSWER IS: (A)

507. Reference: Peurifoy and Oberlender, *Estimating Construction Costs,* 5th ed., pp. 252–253.

There are three errors in the table: an obvious math error in the last line (767.53 is 10× too high), a counting error in the first line (60 should be 120 for the two layers of transverse steel), and a length error in the back face bar in which a portion in the footing is neglected (4.5 should be 6.25). If all errors are caught (corrections shown in bold), the revised table is:

Quantity Take-off to Check				
Item	Number	Length (ft)	Unit Weight (lb/ft)	Weight (lb)
Transverse #6	**120**	5.5	1.502	**991.32**
Longit #4	14	29.5	0.668	275.88
BF Wall #7	119	**6.25**	2.044	**1,520.23**
FF Wall #4	30	3.83	0.668	**76.75**
Total weight				**2,864.18**

THE CORRECT ANSWER IS: (D)

508. References: Frank R. Walker Company, *Walker's Building Estimator's Reference Book*, 28th ed., p. 296; and Halpin and Woodhead, *Construction Management*, 2nd ed., Section 12.8.

Bottom 8 ft $= (40 + 90)(2)(8)(2) = 4,160 \text{ ft}^2 / 48 \text{ ft}^2 / \text{crew-hour} =$ 86.67 crew-hours

Top 6 ft $= (40 + 90)(2)(6)(2) = 3,120 \text{ ft}^2 / (48)(0.6) \text{ ft}^2 / \text{crew-hour} =$ 108.33 crew-hours

                                                      195.0 crew-hours

(195 crew-hours)($372/crew-hour) = $72,540

THE CORRECT ANSWER IS: (D)

CONSTRUCTION PM SOLUTIONS

509. Reference: Peurifoy and Oberlender, *Estimating Construction Costs*, 6th ed., p. 256.

$$\text{Contact area} = 48 \times \frac{12 + 24 + 24}{12} \times 30 = 7,200 \text{ SFCA}$$

Note: Forms are used on sides and bottoms. The top is open, and no forms on ends.

$$\text{Plywood} = \frac{7,200}{100} \times 40 \times 1.00 = \$2,880.00$$

$$\text{Lumber} = \frac{7,200}{100} \times 110 \times 0.90 = \$7,128.00$$

$$\text{Form ties} = \frac{7,200}{100} \times 12 \times 0.60 = \$518.40$$

$$\text{Nails} = \frac{7,200}{100} \times 1.5 \times 1.10 = \$118.80$$

$$\text{Total} \quad\quad\quad\quad\quad\quad\quad\quad\quad\quad \$10,645.20$$

THE CORRECT ANSWER IS: (C)

510. References: Collier and Ledbetter, *Engineering Cost Analysis,* 1982, pp. 275–307; and Smith, *Engineering Economy,* 4th ed., pp. 236–252.

Benefit = reduction in annual costs

$$= 250,000 - 248,000 = \$2,000$$

$$\text{Costs} = 9,000 \left(A/P\right)_{5\,yr}^{10\%} + 1,000 \left(0.10\right)$$

$$= 2,374 + 100 = \$2,474$$

Alternatively,

$$\text{Costs} = 10,000 \left(A/P\right)_{5\,yr}^{10\%} - 1,000 \left(A/F\right)_{5\,yr}^{10\%}$$

$$= 2,638 - 164 = \$2,474$$

$$\text{B/C} = \frac{2,000}{2,474} = 0.81$$

THE CORRECT ANSWER IS: (B)

511. Reference: Newnan, Lavelle, and Eschenbach, *Engineering Economic Analysis*, 10th ed., pp. 298–301.

$$P = F(1 + i)^{-n} \quad \text{and} \quad P = A\left[\frac{(1 + i)^n - 1}{i(1 + i)^n}\right]$$

Purchase price $= 600(4.40) = \$2,640$

Salvage value $= 600(1.00)(1 - 0.25)^{n-1} = 600(0.75)^{n-1}$

Rental cost $= 600(1.50) = \$900/\text{yr}$

Equate present values to find break-even point:

$$2,640 - 600(0.75)^{n-1}(1 + 0.04)^{-n} = 900\left[\frac{(1 + 0.04)^n - 1}{0.04(1 + 0.04)^n}\right]$$

$$0 = 2,640(1.04)^n - 600(0.75)^{n-1} - 22,500(1.04)^n + 22,500$$

$$0 = -33.1(1.04)^n - (0.75)^{n-1} + 37.5$$

Solve for n:

$$\therefore n = 2.77 \text{ years}$$

THE CORRECT ANSWER IS: (B)

512. Reference: Peurifoy and Oberlender, *Estimating Construction Costs*, 5th ed., p. 124.

Total quantity $= 25 \text{ yd}^3$

Total budget $= \$4,000$

Hourly labor cost:

2 at \$25/hr =	\$50/hr
3 at \$35/hr =	\$105/hr
Total =	\$155/hr

$$\text{Total time} = \frac{\$4,000}{\$155/\text{hr}} = 25.8 \text{ hr}$$

$$\text{Required crew productivity rate} = \frac{25 \text{ yd}^3}{25.8 \text{ hr}} = 0.97 \text{ yd}^3/\text{hr}$$

THE CORRECT ANSWER IS: (A)

513. Reference: Meriam and Kraige, *Engineering Mechanics,* vol. 1, Statics, 6th ed., Chapter 3.

The center of gravity of the load is offset. The load is therefore heavier on the left side of the rigging. The portion of the 60-kip load on the left side is (40 ft/60 ft)(60 kips) = 40 kips. The load is further amplified by the slope of the sling. The vector length of Sling A is $\sqrt{50^2 + 20^2} = 53.85$ ft.

The force in the sling is (53.85 ft/50.00 ft)(40 kips) = 43.1 kips.

THE CORRECT ANSWER IS: (D)

514. Reference: Shapiro, Shapiro, and Shapiro, *Cranes and Derricks,* 3rd ed., pp. 361–362.

$$1.5\,M_{OT} = M_{ST}$$

Sum moments at toe of foundation.

Foundation weight $= 16\text{ ft} \times 16\text{ ft} \times 8\text{ ft} \times 150\text{ pcf} = 307.2$ kips

Overturning moment, $M_{OT} = P_p(52\text{ ft}) + (5,000\text{ lb})(7\text{ ft}) + (900\text{ lb})(78\text{ ft}) + (40\text{ lb/ft} \times 80\text{ ft})(48\text{ ft})$

Stabilizing moment, $M_{ST} = (10,000\text{ lb})(38\text{ ft}) + (307,200\text{ lb})(8\text{ ft}) + (7,000\text{ lb})(8\text{ ft})$

$1.5(52\,P_p + 258.8) = 2,893.6$ ft-kips

$78\,P_p = 2,505.4$ ft-kips

$P_p = 32.1$ kips

THE CORRECT ANSWER IS: (B)

515. Reference: Merritt, *Standard Handbook for Civil Engineers,* 3rd ed., pp. 7-94, 7-96.

Wellpoint systems are most effective in permeable soils. The theoretical drawdown limit of a single-stage wellpoint system is about 32 ft, and the practical limit is about 20 ft.

THE CORRECT ANSWER IS: (D)

516. Reference: Allen, *Fundamentals of Building Construction Materials and Methods*, 5th ed., p. 50.

A watertight barrier is the only method that is acceptable. Any drawdown of the water table could result in settlement of the historic masonry structure. Driving sheet piles is likely to cause vibrations that could damage the structure.

THE CORRECT ANSWER IS: (A)

517. Reference: Peurifoy, Schexnayder, and Shapira, *Construction Planning, Equipment and Methods*, 7th ed., Chapter 10.

This operation is weight limited.

Bank density = 110 pcf × 27 ft^3/yd^3 = 2,970 lb/yd^3

Truck capacity (by weight) = $\dfrac{27,000}{2,970}$ = 9.09 yd^3

Trips per 55 min/hr = $\dfrac{55}{17}$ = 3.24

Haulage rate = 3.24 trips/hr × 9.09 yd^3/trip = 29.45 yd^3/hr

THE CORRECT ANSWER IS: (D)

518. References: Cheney and Chassie, *Soils and Foundations Workshop Manual*, FHWA, 1993; and Frank R. Walker Company, *Walker's Building Estimator's Reference Book*, 28th ed., pp. 240–259.

The existing ground elevation from the boring log is Elevation 990. This puts the rock elevation at 929.5 ft. If the bottom of the pile cap is at Elevation 980, the piles need to be 50.5 ft long, plus embedment in the pile cap, plus an allowance for damage during driving.

50.5 + 1.0 + 2.0 = 53.5 ft

THE CORRECT ANSWER IS: (B)

519. Reference: Peurifoy, Schexnayder, and Shapira, *Construction Planning, Equipment and Methods,* 7th ed.

Gross load for Package 1 = 17,500 + 8,000 = 25,500 lb. Referring to the table, this is between the entries for 110 ft (23,900 lb) and 105 ft (25,800 lb), so choose 105 ft. Locate building edge on graph at a distance of 60 ft and a height of 90 ft and reduce distance to 53 ft to allow for boom clearance. This places the boom centerline about on the 65° angle. For the 180 ft boom, the reach is limited to 90 ft. However, for load radii greater than 90 ft, the building interferes with the boom, so max load radius is 90 ft. Thus, 90 − 60 = 30 ft, distance from edge of roof.

Gross load for Package 2 = 39,200 + 8,000 = 47,200 lb. Referring to the table, this is between the entries for 70 ft (47,600 lb) and 75 ft (43,100 lb), so choose 70 ft. Thus, 70 − 60 = 10 ft from edge of roof.

THE CORRECT ANSWER IS: (D)

520. Reference: Gould, *Managing the Construction Process,* 2nd ed., pp. 247–249.

Option (A) meets all the successor requirements.

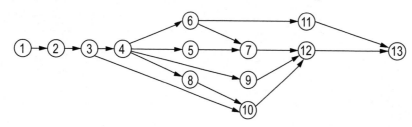

THE CORRECT ANSWER IS: (A)

521. Reference: Schexnayder and Mayo, *Construction Management Fundamentals,* 2004, pp. 496–502.

$$\text{Duration} = 420 \text{ fixtures} / \left[\left(\frac{1 \text{ fixture}}{20 \text{ min} \times 2 \text{ electricians}} \right) \times 6 \text{ electricians} \times 0.8 \times 60 \text{ min/hr} \times 8 \text{ hr/day} \right]$$
$$= 7.29 \text{ workdays}$$

THE CORRECT ANSWER IS: (C)

522. Reference: Oberlender, *Project Management for Engineering and Construction*, 2nd ed., Chapter 8, pp. 139–184.

Network calculations are shown below. Since FF in E ≥ 2 days, delay will not affect any other activity.

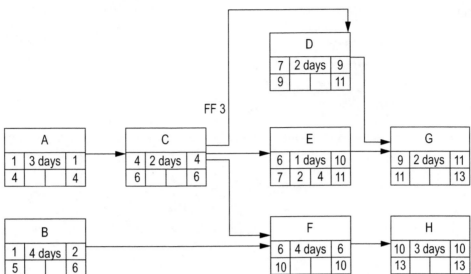

THE CORRECT ANSWER IS: (A)

523. Reference: Oberlender, *Project Management for Engineering and Construction*, 2nd ed., pp. 164–167.

This result can be achieved by inspection or by using the improvement factor by first moment. It requires shifting Activity F 2 days later, from ES to LS, and then shifting Activity D 2 days later, from ES to LS. The final resource calculation is:

Activity	Period					
	1	**2**	**3**	**4**	**5**	**6**
A	6					
B		7	7			
C		2	2	2	2	2
D				2		
E				3		
F					8	8
Total	6	9	9	7	10	10

THE CORRECT ANSWER IS: (D)

524. Reference: Hinze, *Construction Planning and Scheduling*, 2nd ed., 2004, Chapter 7.

Straight-time cost:

4,000 L-H × $10/L-H = $40,000

General conditions:

40 hr/week × 10 workers = 400 L-H/week

4,000 L-H/400 L-H/week = 10 weeks = 50 days

50 days × $1,000/day = $50,000

Total cost = $40,000 + $50,000 = $90,000

Least duration is 9 weeks due to subcontracts (subcontractors stay off critical path)

Overtime needed at 9 weeks:

4,000 L-H/9 weeks × 10 workers = 44.44 hr/week

Labor cost:

Straight time	40 hr/week × 10 workers × 9 weeks × $10/L-H =	$36,000
Overtime	4.44 hr/week × 10 workers × 9 weeks × $15/L-H =	$6,000
		$42,000
General conditions: 45 days × $1,000/day		$45,000
Total		$87,000

Savings = $90,000 − 87,000 = $3,000

THE CORRECT ANSWER IS: (A)

525. Reference: Bowles, *Foundation Analysis and Design*, 5th ed., pp. 32–33.

The sample has the following properties:

% passing the #4 sieve	77%
% fines passing the #200 sieve	18%
% retained on the #200 sieve	100% − 18% = 82%
Liquid Limit, LL	32
Plastic Limit, PL	25
Plasticity Index, PI = LL − PL	7

Based on 77% passing the #4 sieve and 82% retained on the #200 sieve, the soil is classified as a sand, either SM or SC. Based on the fines having LL = 32 and PI = 7, the fines would be classified as ML, nonplastic. Therefore, according to the Unified Soil Classification System, the sample is classified as SM.

THE CORRECT ANSWER IS: (C)

526. Reference: Gere, *Mechanics of Materials*, 6th ed., 2006, pp. 68–76.

P = 14 tons = 28 kips

$\Delta = PL/(A_{ps}E_{ps}) = 28$ kips $(310 \text{ ft})(12 \text{ in./ft})/(0.153 \text{ in}^2 \times 27,500 \text{ ksi}) = 24.7$ in.

THE CORRECT ANSWER IS: (D)

527. Reference: AISC *Steel Construction Manual*, 14th ed., p. 8-35, Table 8-2.

The symbol indicates a field-placed bevel weld. Weld symbol placed above line indicates weld on the other side (side opposite the arrow).

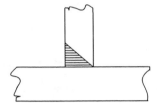

THE CORRECT ANSWER IS: (D)

528. Reference: ACI 315-99: "Details and Detailing of Concrete Reinforcement," *ACI Detailing Manual*, 2004.

Correct interpretation involves #8 bar size, 6-in. spacing, and alternating projection above finish floor elevation.

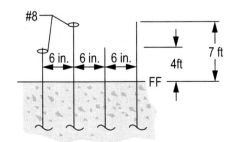

THE CORRECT ANSWER IS: (A)

529. Reference: Kosmatka, Kerkhoff, and Panarese, *Design and Control of Concrete Mixtures,* PCA, 14th ed., Chapter 9.

Work on the basis of 1 yd^3 of concrete.

Weight of cement $= 590$ lb/yd^3

Volume of cement $= (590 \text{ lb/yd}^3)(1/3.15)(1 \text{ ft}^3/62.4 \text{ lb}) = 3.00 \text{ ft}^3/\text{yd}^3$

Vol of fine aggregate $= (2.25)(590 \text{ lb/yd}^3)(1/2.65)(1 \text{ ft}^3/62.4 \text{ lb}) = 8.03 \text{ ft}^3/\text{yd}^3$

Vol of coarse aggregate $= (3.25)(590 \text{ lb/yd}^3)(1/2.65)(1 \text{ ft}^3/62.4 \text{ lb}) = 11.60 \text{ ft}^3/\text{yd}^3$

Vol of water/yd^3 $= 27.0 - 3.00 - 8.03 - 11.6 = 4.37 \text{ ft}^3$

Weight of water $= (4.37 \text{ ft}^3)(62.4 \text{ lb/ft}^3) = 272.69 \text{ lb}$

Water/cement ratio $= 272.69 \text{ lb}/590 \text{ lb} = 0.46$

THE CORRECT ANSWER IS: (B)

530. Reference: Kosmatka, Kerkhoff, and Panarese, *Design and Control of Concrete Mixtures,* PCA, 14th ed., p. 254.

The concrete maturity, $M = \Sigma(T - T_0)\Delta t$ where $\Delta t = 4$ hours

Summation of temperatures minus the datum temperature $(T - 20°F)$:

$46 + 50 + 56 + 60 + 62 + 62 + 58 + 54 + 52 + 48 + 44 + 40 = 632$

$M = 632°F \times 4 \text{ hr} = 2{,}528°F\text{-hr}$

THE CORRECT ANSWER IS: (C)

531. Reference: Hibbeler, *Engineering Mechanics: Statics,* 12th ed., 2010, pp. 494–504.

Concrete density under water	$150 - 62.4$	$= 87.6 \text{ lb/ft}^3$
Block weight under water	$(60 \times 20 \times 4)(87.6)$	$= 420{,}480 \text{ lb}$
Contact area	60×20	$= 1{,}200 \text{ ft}^2$
Contact pressure	$\dfrac{420{,}480}{1{,}200}$	$= 350 \text{ lb/ft}^2$

THE CORRECT ANSWER IS: (A)

532. Reference: ACI 347-04, *Guide to Formwork for Concrete*, 2004, Section 2.2.2.

For placement rates of 7–15 ft/hr:

$$p_{max} = C_W C_C \left[150 + \frac{43{,}400}{T} + 2{,}800 \, R/T \right]$$

with a minimum of 600 (C_W), but in no case greater than wh

$$p_{max} = \left(\frac{155}{145} \right)(1.2) \left[150 + \frac{43{,}400}{80} + 2{,}800(15)/(80) \right]$$

$$= (1.069)(1.2)(1{,}217.5) = 1{,}561 \text{ lb/ft}^2$$

Check minimum $= 600 \, C_W$

$\qquad\qquad\quad = 600 \times 1.07$

$\qquad\qquad\quad = 641 \text{ lb/ft}^2$

Check maximum $= wh$

$\qquad\qquad\quad = 155 \times 9$

$\qquad\qquad\quad = 1{,}395 \text{ lb/ft}^2$

Therefore use the maximum of 1,395 lb/ft^2 (cannot exceed a liquid head)

THE CORRECT ANSWER IS: (B)

533. Reference: OSHA 29 CFR 1926.451(c) (1) (ii) of Subpart L.

OSHA 1926 451(c) (1) (ii) requires that scaffolds 4 ft and wider be braced at the closest horizontal member to the 4 × width height and be repeated vertically every 26 ft or less at locations of horizontal members. The top brace shall be no further than the 4:1 height from the top.

Maximum height from ground = 4 × 5 ft = 20 ft
Maximum distance from top = 4 × 5 ft = 20 ft
Maximum distance between braces = 26 ft

Option (C) satisfies these requirements.

THE CORRECT ANSWER IS: (C)

534. References: ACI SP-4, *Formwork for Concrete*, 7th ed., 2005, Chapter 5, Table 5.3, 5.3A; and ACI 347.2R-05, *Guide for Shoring/Reshoring of Concrete Multistory Buildings*, 2005.

Load	Net change
Carried by Floors 7 and 8 with shore removal from under Level 7	$(1.50\ D - 0.10\ D)/2$ levels $= 0.70\ D$ each
Carried by Floors 7 and 8 due to form storage on Level 8	$0.10\ D/2$ levels $= 0.05\ D$ each
Carried by Floors 7 and 8 due to worker live load on Level 8	$0.40\ D/2$ levels $= 0.20\ D$ each

The load carried by Floor 7 as operation indicated is nearing its end is:

Existing load + added load due to shore removal + added load due to stored forms + added load due to worker live load

$= 0.70\ D + 0.70\ D + 0.05\ D + 0.20\ D$

$= 1.65\ D$

THE CORRECT ANSWER IS: (C)

535. Reference: Meriam and Kraige, *Engineering Mechanics,* vol. 1, Statics, 6th ed., Chapter 3.

$\Sigma M_A = 0$

$(12 \text{ kips})(3 \text{ in.}) - T(8 \text{ in.}) = 0$

$T = \dfrac{(12 \text{ kips})(3 \text{ in.})}{8 \text{ in.}}$

$= 4.5 \text{ kips}$

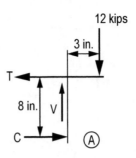

FREE-BODY DIAGRAM

THE CORRECT ANSWER IS: (B)

536. Reference: OSHA 29 CFR 1926, *Safety and Health Regulations for Construction*, Subpart P, Excavations.

For the special case of a Type A soil with excavation not exceeding 12 ft and open for the short term of 24 hours or less, a maximum slope of 1/2 horizontal to 1 vertical is allowed.

THE CORRECT ANSWER IS: (B)

537. Reference: ACI SP-4: *Formwork for Concrete*, M.K. Hurd, ACI, 7th ed., Table 6-1, p. 5-1.

Dead load : $\dfrac{6 \text{ in.}}{12 \text{ in./ft}} \times 150 \text{ lb/ft}^3 = 75 \text{ lb/ft}^2$

Live load $\qquad\qquad\qquad \underline{50 \text{ lb/ft}^2}$

Total load $\qquad\qquad\quad 125 \text{ lb/ft}^2$

Load on stringer

$W = 125 \text{ lb/ft}^2 (3 \text{ ft}) = 375 \text{ lb/ft}$

Reaction at B2

$R = 1.25 \text{ WL} = 1.25 (375 \text{ lb/ft})(5 \text{ ft}) = 2{,}344 \text{ lb}$

THE CORRECT ANSWER IS: (C)

538. Reference: OSHA 29 CFR 1926.62, Subpart D (f) (3) (i), Table 1.

Half-mask or full-facepiece supplied-air respirator operated in continuous-flow mode. 1,500 μg/m^3 is more than 1,250 μg/m^3 and less than 2,500 μg/m^3.

THE CORRECT ANSWER IS: (D)

539. Reference: OSHA Forms 300 and 300A.

The term *incidence rate* for recordable cases means the number of injuries and illnesses requiring treatment beyond first aid, those that result in lost workdays, and those that result in restricted or light duty, per 100 full-time workers. The rate is calculated as

N × 200,000/EH

where

N = number of injuries and illnesses, or number of lost workdays.

EH = total hours worked by all employees during a month, a quarter, or a fiscal year.

200,000 = base for 100 full-time equivalent workers (including 40 hours per week, 50 weeks per year).

In this case N = 4 + 3 + 5 = 12

The incidence rate = 12 (200,000)/750,000 = 3.20

THE CORRECT ANSWER IS: (C)

540. Reference: FHWA, *Manual of Uniform Traffic Control Devices*, 2009, p. 557.

Using Table 6C-4, taper length L:

$$L = \frac{WS^2}{60} = \frac{12(35)^2}{60} = 245 \text{ ft}$$

Referring to Table 6C-3, merging taper (minimum) = L

THE CORRECT ANSWER IS: (C)

PE Practice Exams Published by NCEES

Chemical

Civil: Geotechnical

Civil: Structural

Civil: Transportation

Civil: Water Resources and Environmental

Electrical and Computer: Electrical and Electronics

Electrical and Computer: Computer Engineering

Electrical and Computer: Power

Environmental

Mechanical: HVAC and Refrigeration

Mechanical: Mechanical Systems and Materials

Mechanical: Thermal and Fluids Systems

Structural

For more information about these and other NCEES publications and services,
visit NCEES.org or call Client Services at (800) 250-3196.